TABLES OF
ELECTRIC DIPOLE MOMENTS

COMPILED BY

L. G. WESSON

Laboratory for Insulation Research
Massachusetts Institute of Technology
Cambridge, Massachusetts

A PUBLICATION OF

THE TECHNOLOGY PRESS

MASSACHUSETTS INSTITUTE OF TECHNOLOGY

The compilation of the tables in this book was made possible through support extended to the Laboratory for Insulation Research, Massachusetts Institute of Technology, jointly by the Navy Department (Office of Naval Research) and the Army Signal Corps under ONR Contract N5ori-78, T. O. 1.

Preface

Since Debye[1] postulated the existence of permanent dipole moments, this molecular constant has become of decisive importance for the interpretation of the dielectric properties of gases, liquids and solids. More than twenty-five hundred compounds have been measured to date, many under a variety of conditions and by a number of different methods. These substances were selected according to the diverse interests of the investigators, and the results are scattered in the physical and chemical literature of many countries.

We felt the need, from the standpoint of coordinated research on dielectrics, for a well-organized compilation of these dipole moments[2], and L. G. Wesson of this Laboratory has undertaken the arduous and exacting task.

It is hoped that the "Tables of Electric Dipole Moments" will prove of value to all research workers in this field, and that we may count on the support of previous, as well as future authors, in making these "Tables" more reliable, complete and up-to-date.

Cambridge, Mass.
August, 1948

A. von Hippel, Director
Laboratory for Insulation
Research

1) P. Debye, Physik. Z. 13, 97 (1912).
2) For earlier summaries see:- C. P. Smyth, Dielectric Constant and Molecular Structure, Chemical Catalog Co., New York, 1931; P. Debye, Polare Molekeln, Hirzel, Leipzig, 1929, with supplementary lists dated 1930 and 1931; N. V. Sidgwick, G. C. Hampson and R. J. B. Marsden, Trans. Far. Soc. 30, Appendix (1934); O. Fuchs and K. L. Wolf, Hand- und Jahrbuch der chemischen Physik 6, I (1935); Physikalisch-chemische Tabellen, Landolt-Börnstein, Springer, Berlin, 5th Ed., Suppl. II a, pp. 74-86; II b, pp. 969, 970; III a, pp. 117-150 (1935); R. J. W. Le Fèvre, Dipole Moments, Methuen and Co. Ltd., London (1938).

TABLES OF
ELECTRIC DIPOLE MOMENTS

Substances for which electric dipole moments have been found in the literature are presented in these tables in three groups:

    I.   Inorganic substances (pages 2 to 4)
    II. Metal-organic, silicon-organic, and boron-organic compounds (pages 4 to 11)
    III. Organic compounds (pages 12 to 64).

The arrangement of I and II is alphabetical, while III is organized according to the empirical formulas.

The dipole moments (underlined) are given in Debye units (1 Debye unit = 1 x 10^{-18} electrostatic units).

The temperatures at which gases or vapors have been measured, or to which values are extrapolated or interpolated, are given in degrees absolute (^{O}K); reference to the temperature in degrees absolute indicates that the substance has been measured in the form of a gas or vapor. Often the temperature has not been stated and "Gas" or "Vap." is used to indicate the state of aggregation in these cases. For solids, liquids or solutions of these, the reference temperatures are given in degrees centigrade, and measurements in the liquid or solid state are indicated by "Lqd." or "Solid."

Frequently-used solvents are shown by the following abbreviations:

Am = amylene	Dec = decalin ($C_{10}H_{18}$)
An = aniline	E = ethyl ether
B = benzene	Hp = n-heptane
CB = chlorobenzene	Hx = n-hexane
CD = carbon disulfide	M = mesitylene
Cf = chloroform	N = naphthalene
CH = cyclohexane	NB = nitrobenzene
CT = carbon tetrachloride	T = toluene
D = 1,4-dioxane	pX = p-xylene.

If the solvent is not given by the author the letter "S" is used.

In general the dipole values for a substance are arranged in chronological order starting with the earliest reference. Numbers enclosed in parentheses following the dipole values refer to the bibliography, pages 65 to 83.

An author index to the bibliography concludes the volume.

I. Inorganic Substances

Air, 392-444°K $\underline{0}$ (1); 87°K $\underline{0.026}$ (2); 290°K $\underline{0}$ (18); 291-457°K ca. $\underline{0}$ (56); 298°K $\underline{0}$ (394).

Aluminum bromide, B $\underline{4.89}$, CD $\underline{0.66}$ (195); 20° B $\underline{5.2}$, CD $\underline{0.55}$ (247); B $\underline{5.03}$, CD $\underline{0}$, bromine $\underline{0}$ (616, 659).

Aluminum bromide-hydrogen sulfide [AlBr$_3$.H$_2$S], B $\underline{5.14}$ (194); B $\underline{5.11}$ (195); B 20° $\underline{5.25}$ (247).

Aluminum chloride, D 25° $\underline{2.02}$ ± .05 (765).

Aluminum iodide, 20° B $\underline{2.5}$, CD $\underline{0}$ (247).

Ammonia, 292-448°K $\underline{1.53}$ ±.01 (1); 242-457°K $\underline{1.44}$ (21); 226-289°K $\underline{1.51}$, 289-373°K $\underline{1.46}$ (29); 373-448°K $\underline{1.44}$ (123); 373-473°K $\underline{1.48}$ (325); 274-423° K $\underline{1.46}_6$ (407); 25° B $\underline{1.38}$, Hp $\underline{1.43}$ (489); Gas $\underline{1.3}$ (717); 298°K $\underline{1.45}$, B 25° $\underline{1.40}$, Lqd. 5-35° $\underline{0.93}$ (852).

Antimony pentachloride, CT 16° $\underline{1.14}$ (152-154); CT 7-47° ca. $\underline{0}$ (179).

Antimony tribromide, S $\underline{2.4}$ (309); 25° B $\underline{3.28}$ ± .1, CD $\underline{2.47}$ ± .1 (365); D 25° $\underline{5.01}$ (767).

Antimony trichloride, B $\underline{3.64}$, E $\underline{3.90}$ (96); B $\underline{4.06}$ (152); B 16° $\underline{4.11}$ (153, 154); B $\underline{3.75}$, B 25° $\underline{3.75}$ (255); S $\underline{3.1}$ (309); CD 25° $\underline{3.12}$ ± .05 (365); D 25° $\underline{5.16}$ (767).

Antimony triiodide, B 25° $\underline{0.4}$ (61); CD 25° $\underline{1.58}$ ±.1 (365).

Argon, 296-383°K $\underline{0}$ (22); 273-383°K $\underline{0}$ (151); 63-298°K < $\underline{0.03}$ (197).

Arsenic tribromide, CT 18° $\underline{1.66}$ (152, 153, 154); 25° B $\underline{3.28}$ ± .1, CD $\underline{2.47}$ ± .1 (365); D 25° $\underline{2.90}$ (767).

Arsenic trichloride, CT 18° $\underline{1.97}$ (152-154); B $\underline{2.15}$, B 25° $\underline{2.17}$ (255); 403-468°K $\underline{1.6}$ (535); D 25° $\underline{3.11}$ (767).

Arsenic trifluoride, B 25° $\underline{2.65}$ ± .05 (365).

Arsenic triiodide, CD $\underline{0.96}$ ± .1 (365); D 25° $\underline{1.83}$ (767).

Arsenic trioxide, 525°K $\underline{0.13}$ (70).

Arsine, 226-289°K $\underline{0.13}$, 289-373°K $\underline{0.18}$ (29); Solid -239 to -193°C $\underline{0.15}$ (508).

Beryllium bromide, B ca. $\underline{0}$ (195).

Beryllium chloride, B ca. $\underline{0}$ (195).

Boron bromide, Solid and Lqd. -70 to + 80° ca. $\underline{0}$ (833a).

Boron chloride, CT 18° $\underline{0.21}$ (152-154); B $\underline{0}$ (195); B 20° $\underline{0}$ (247); 308-442°K $\underline{0.60}$ (535); D 25° $\underline{4.86}$ ± .07 (765).

Boron fluoride, 193-298°K $\underline{0}$ (514); 293-472°K $\underline{0}$ (555).

Boron hydride (diborane) [B$_2$H$_6$] , 153-368°K $\underline{0}$ (445).

Bromine, Lqd. 0-20° $\underline{0.40}$ ±.04 (33); Lqd. 0-54° $\underline{0.49}$ (107); 293-412°K < $\underline{0.1}$ (307).

Cesium chloride, 928°K ca. $\underline{10}$ (32).

Chlorine, Lqd. -65 to 8° $\underline{0.23}$ (151).

Chlorine dioxide, CT -18 to 24° $\underline{0.78}$ ± .08 (452).

Chlorine heptoxide, CT $\underline{0.72}$ (587).

Chlorine monoxide, CT -18 to 24° $\underline{1.69}$ ± .09 (452).

Chromyl chloride, CT 25° $\underline{0.47}$ (702).

Deuteroammonia, 274-425°K $\underline{1.49}_6$ (407).

Deuterium chloride, 291-517°K $\underline{1.08}_9$ (575).

Deuterium oxide, B 20° $\underline{1.78}$ ± .02 (367); D 20° $\underline{1.87}$ (400); 364-473°K $\underline{1.84}$ (417).

Ferric chloride, D 25° $\underline{1.27}$ ± .05 (765).

Gallium trichloride, CD R.T. $\underline{0.8}$ (744).

Germane, dichloro-, CT 25° $\underline{2.21}$ (651).

Germane, monochloro-, 276-297°K $\underline{2.03}$ (702).

Germanium tetrachloride, D 25° $\underline{0.67}$ ± .08 (765).

Helium, Lqd. -271 to -269°C $\underline{0}$, 273°K $\underline{0}$ (151); 63-298°K < $\underline{0.015}$ (197).

Hydrazine, B 18° $\underline{1.83}$ to $\underline{1.85}$ (275).

Hydrogen, 82°K $\underline{0.043}$ (2); 289-383°K $\underline{0}$ (22); 295-571°K $\underline{0.038}$ + .016 (43); Lqd. -259 to -253°C $\underline{0}$, 289-571°K $\underline{0}$ (151); 63-298°K < $\underline{0.015}$ (197);

273-373°K $\underline{0}$ (325).

Hydrogen bromide, 206-673°K $\underline{0.7881}$ (10); 218-599°K $\underline{0.79}$ (11); 20° B $\underline{1.01}$, CT $\underline{0.96}$ (347); B $\underline{1.5}$, CT $\underline{0.6}$ (550); B $\underline{1.08}$, Hp $\underline{1.02}$, CT $\underline{0.93}$, D $\underline{2.85}$ (870).

Hydrogen chloride, Gas $\underline{2.15}$ (6); 293°K $\underline{1.48}$ (9); 189-673°K $\underline{1.03}$ (10); 201-589°K $\underline{1.034}$ (11); 289-389°K $\underline{1.180}$ ± .05 (22); Kerr 293°K $\underline{1.04}$ (25); 25° B $\underline{1.28}$, CH $\underline{1.32}$, CT $\underline{1.32}$ (227); Gas ca. $\underline{1.95}$ (286); 20° B $\underline{1.26}$, ethyl bromide $\underline{1.02}$, ethylene chloride $\underline{0.97}$ (288); B 20° $\underline{1.26}$ (347); B 10-25° $\underline{1.21}$, T 20° $\underline{1.24}$, Hx 20° $\underline{1.04}$, E 30° $\underline{2.22}$, Cf 30° $\underline{1.03}$ (434); Gas $\underline{1.18}$ (447); B $\underline{3.0}$, CT $\underline{1.6}$, CB $\underline{1.6}$ (550); 291-517°K $\underline{1.08}_2$ (575); B $\underline{1.25}$, Hp $\underline{1.19}$, CT $\underline{1.06}$, D $\underline{2.12}$ (870).

Hydrogen fluoride, 305-374°K $\underline{1.91}$ (828); 30° B $\underline{1.91}$, Hp $\underline{2.2}$, CT $\underline{2.0}$, D $\underline{2.34}$ (870).

Hydrogen iodide, 245-612°K $\underline{0.38}$ (11); 20° B $\underline{0.58}$, CT $\underline{0.50}$ (347).

Hydrogen peroxide, 25° D $\underline{2.13}$, E $\underline{2.06}$ (242, 324); Lqd. $\underline{2.13}$ (562).

Hydrogen sulfide, 287-387°K $\underline{1.101}$ ± .05 (22); 197-542°K $\underline{0.931}$ (67); B 25° $\underline{1.17}$ (703).

Iodine, B 25° $\underline{1.4}$ (30); E 20° $\underline{0.7}$ (298,358); B 15-70° $\underline{0}$, CD 15-35° $\underline{0}$ ±.1 (458).

Iodine monochloride, 334-435°K ca. $\underline{0.5}$ (307); CT 25° $\underline{0.9}$ to $\underline{1.1}$ (365); CH $\underline{1.47}$, CT $\underline{1.49}$ (470).

Krypton, 63-298°K < $\underline{0.03}$ (197).

Lithium perchlorate, D 25° $\underline{7.84}$ ± .05 (365).

Mercuric bromide, D 25° $\underline{1.06}$ (413); D 25° $\underline{1.53}$ (715).

Mercuric chloride, D 25° $\underline{1.29}$ (413); D 25° $\underline{1.43}$ (715).

Mercuric iodide, D 25° $\underline{0.58}$ (413); D 25° $\underline{1.67}$ (715).

Neon, 273°K $\underline{0}$ (151); 63-298°K < $\underline{0.015}$ (197).

Nitric oxide, 235-335°K $\underline{0.0}_7$ (318); 298°K $\underline{0.16}$ (394).

Nitroamine, D 20° $\underline{3.75}$ (303).

Nitrogen, 290°K $\underline{0}$ (18); 84-562°K $\underline{0}$ (21); 295-583°K $\underline{0.06}$ ±.01 (43); Lqd. 64-77°K $\underline{0}$, 84-562°K $\underline{0}$ (151); 273-373°K $\underline{0}$ (325); 298°K $\underline{0}$ (394).

Nitrogen dioxide, 297-397°K $\underline{0.3}_3$ (331); 325-363°K $\underline{0.29}$ (621).

Nitrogen pentoxide, CT 25° $\underline{1.39}$ (652).

Nitrogen tetroxide, 297-397°K $\underline{0.4}_7$ (331); 298-398°K $\underline{0}$ (397); 325-363°K $\underline{0.37}$ (621).

Nitrogen tetroxide ⇌ nitrogen dioxide, 298°K $\underline{0.58}$, 318° $\underline{0.49}$, 343° $\underline{0.41}$, 368° $\underline{0.35}$, 398° $\underline{0.30}$ (515).

Nitrogen trifluoride, 193-368°K $\underline{0.21}$ (445); 193-298°K $\underline{0.25}$, 298-368°K $\underline{0.21}$ (514).

Nitrosyl bromide, CT 12° $\underline{1.87}_5$ (788).

Nitrosyl chloride, CT 12° $\underline{1.83}$ ± .01 (788).

Nitrous oxide, 288-369°K $\underline{0.249}$ ± .05 (22); 305-413°K $\underline{0}$ (75); 301-426°K $\underline{0.05}$ (135); 235-335°K $\underline{0.0}_7$ (318); 293-454°K $\underline{0.14}$ ± .02 (344); 298°K $\underline{0.17}$ (394).

Osmuim tetroxide, B $\underline{0}$ (370).

Oxygen, 290°K $\underline{0}$ (18); Lqd. -218 to -183° $\underline{0}$, 273°K $\underline{0}$ (151); 298°K $\underline{0}$ (394).

Ozone, liquid oxygen -192° $\underline{0.49}$ (651).

Phosphine, 226-289°K $\underline{0.54}$, 289-373°K $\underline{0.56}$ (29).

Phosphorus, Lqd. 40-80° $\underline{0}$ (285).

Phosphorus oxychloride, B 20° $\underline{2.40}$ (498); B 10° $\underline{2.36}$, 25° $\underline{2.39}$, 40° $\underline{2.41}$, 60° $\underline{2.42}$ (703).

Phosphorus pentachloride, CT 9-32° ca. $\underline{0}$ (179); CT or CD 25° $\underline{0.8}$ (510).

Phosphorus tribromide, CT $\underline{0.61}$ (152, 153); CT 18° $\underline{0.61}$ (154); D 25° $\underline{1.65}$ (767).

Phosphorus trichloride, CT 18° $\underline{0.80}$ (152-154); Solid $\underline{0.70}$ (243); B $\underline{0.90}$, B 25° $\underline{0.90}$ (255); D 25° $\underline{1.89}$ (767).

Phosphorus trifluoride, 283-388°K $\underline{0}$ (555).

Phosphorus triiodide, CD $\underline{1.58}$ ± .1 (365).

Potassium, Vap. ca. $\underline{0}$ (70).

Potassium chloride, 1024°K $\underline{6.3}$ (384).

Potassium iodide, 928°K ca. $\underline{10}$ (32); 941-951°K $\underline{6.8}$ (384).

Rubidium bromide, 928°K ca. $\underline{10}$ (32).

Selenium monochloride, B 25° $\underline{2.1}$ (703).

Selenium oxychloride, B 25° $\underline{2.62}$ (703).

Silane $\left[SiH_4\right]$, 298°K $\underline{0}$ (394).
Silane, chloro-, 288-402°K $\underline{1.284}$± .005 (581).
Silane, dichloro-, 291-397°K $\underline{1.173}$ ± .005 (581).
Silane, tribromo- (silicobromoform), Hp 25° $\underline{0.79}$ (651).
Silane, trichloro-, 285-414°K $\underline{0.850}$ ± .005 (581).
Silicoethane $\left[Si_2H_6\right]$, 298°K $\underline{0}$ (394).
Silicon tetrachloride, CT 19° $\underline{0}$ (152-154); D 25° $\underline{0}$ (765).
Silicon tetrafluoride, 298°K $\underline{0}$ (394).
Silver perchlorate, B 25° $\underline{4.7}$ (30); B 25° $\underline{11.97}$ (306); B 25ᵘ $\underline{10.7}$ (359).
Sodium,   Vap. ca. $\underline{0}$ (70); Solid $\underline{3.8}$ (522).
Sodium iodide, 928°K ca. $\underline{10}$ (32); 925-975°K $\underline{4.9}$ (384).
Stannic chloride, CT 18° $\underline{0.80}$ (152-154); B $\underline{0}$ (195); B 15-20° $\underline{0.80}$, CT 15-19° $\underline{0}$,
        Lqd. -20 to 20° $\underline{0}$, Solid -35° $\underline{0}$ (263); Hx or CT 25° $\underline{0}$ (321); D 25° $\underline{3.82}$
        ± .15 (765).
Stannic iodide, B 25° $\underline{0}$ (30).
Sulfamide $\left[(NH_2)_2SO_2\right]$, D 20° $\underline{3.9}$ (284).
Sulfur,   Vap. $\underline{1.87}$ (6); Lqd. and Solid 18-125° $\underline{7.5}$ (28); Lqd. 95-150° $\underline{0}$, B 22-
        67° $\underline{0}$, CD 8-35° $\underline{0}$ (48); B $\underline{0}$, CD ca. $\underline{0}$ (132); CD 17-31° $\underline{0}$ (222); Lqd.
        118-350° $\underline{0}$ (283).
Sulfur dioxide, 291-446°K $\underline{1.76}$ (1); 292-293°K $\underline{1.83}$ (3); 290°K $\underline{1.7}$ (18); 266-444°K
        1.611 (21).
Sulfur hexafluoride, 298ᵁK $\underline{0}$ (394).
Sulfur monochloride, B $\underline{1.06}$ ± .02, CT $\underline{0.92}$ ± .02 (619); B 25° $\underline{1.60}$ (703).
Sulfur nitride $\left[S_4N_4\right]$, B 30° $\underline{0.72}$ (658).
Sulfur trioxide, 193-323°K $\underline{0}$ (565).
Sulfuryl chloride, B 25-45° $\underline{1.64}$ (256); 293-416°K $\underline{1.795}$ + .005 (636).
Tantalum chloride, CD 25° $\underline{1.2}$ (503).
Tellurium tetrachloride, B 25° $\underline{2.54}$ (702); B 25° $\underline{2.57}$ (784).
Thionyl bromide, B 20° $\underline{1.47}$ (579).
Thionyl chloride, B 25-45° $\underline{1.38}$ (256); B 20° $\underline{1.60}$ (579); 288-407°K $\underline{1.44}$ ±.005
        (636).
Thiophosphoryl chloride, B or Hp 25° $\underline{1.41}$ (703).
Titanium tetrachloride, CT 20° $\underline{0}$(152-154); Solid -30° $\underline{0}$, Lqd. -20 to 20° $\underline{0}$, B 14°
        $\underline{0}$, CT 19° $\underline{0}$ (263).
Vanadium pentoxide (vanadic anhydride), Colloidal $\underline{415}$ (36).
Water,   390-451°K $\underline{1.87}$ (1); Lqd. 18° $\underline{0.74}$ (4); 393-483°K $\underline{1.84}_7$ ± .02 (49, 85);
        B 25° $\underline{1.70}$ ± .06 (59, 116); B 25° $\underline{1.7}$ (61); E 17° $\underline{1.97}$ (121); E $\underline{1.9}$ (122);
        25° B $\underline{1.7}_2$, D $\underline{1.9}$ (144); 25° D $\underline{1.90}$, E $\underline{1.71}$ (242); 393-483°K $\underline{1.84}_2$ ±
        .008 (254); E 20° $\underline{1.9}$ (298, 358); B 20° $\underline{1.76}$ ± .02 (367); D 20° $\underline{1.86}$
        (400); 383-484°K $\underline{1.84}$ ± .01 (417); D 15 and 30° $\underline{1.91}$ ± .02 (707); 384-
        522°K $\underline{1.84}$ (758); Lqd. 25° $\underline{3.11}$ (816); Stark $\underline{1.84}$ (864); D $\underline{1.89}$ (870).
Xenon,   63-298°K< $\underline{0.05}$ (197).

II. Metal-Organic Compounds

(including metallic salts of organic acids,
silicon-organic and boron-organic compounds)

Aluminum
    Aluminum acetoacetate $\left[(CH_3COCH_2CO_2-)_3Al\right]$, B $\underline{3.96}$ (746).
    Aluminum bromide·anisole $\left[AlBr_3.C_6H_5OCH_3\right]$, B $\underline{6.58}$ (774).
    Aluminum bromide·benzene $\left[AlBr_3·C_6H_6\right]$, S $\underline{5.1}$ (247).
    Aluminum bromide·benzophenone $\left[AlBr_3·(C_6H_5)_2CO\right]$, B $\underline{8.41}$ (774).

Aluminum bromide·o-chloronitrobenzene [AlBr$_3$·C$_6$H$_4$ClNO$_2$], B 9.56 (774).
Aluminum bromide·diphenylamine [AlBr$_3$·(C$_6$H$_5$)$_2$NH], B 6.68 (774).
Aluminum bromide·ethyl ether [AlBr$_3$·(C$_2$H$_5$)$_2$O], B 6.54 (194); B 6.39
 (195); B 20° 6.59 (247).
Aluminum bromide·nitrobenzene [AlBr$_3$·C$_6$H$_5$NO$_2$], B 9.13 (194); B 9.08
 (195); B 20° 9.30 ± .13 (247).
Aluminum bromide·o-nitrotoluene [AlBr$_3$·CH$_3$C$_6$H$_4$NO$_2$], B 9.30 (774).
Aluminum bromide·p-nitrotoluene [AlBr$_3$·CH$_3$C$_6$H$_4$NO$_2$], B 9.76 (774).
Aluminum bromide·phenyl ether [AlBr$_3$·(C$_6$H$_5$)$_2$O], B 6.56 (774).
Aluminum n-butyrate [(C$_3$H$_7$CO$_2$-)$_3$Al], B 25° 3.24 (371).
Aluminum chloride·anisole [AlCl$_3$·CH$_3$OC$_6$H$_5$], S 6.54 (700).
Aluminum chloride·benzophenone [AlCl$_3$·(C$_6$H$_5$)$_2$CO], B 8.72 (194); B 8.65
 (195); B 20° 8.32 ± .14 (247).
Aluminum chloride·benzoyl chloride [AlCl$_3$·C$_6$H$_5$COCl], B 8.92 (194); B
 8.87 (195); 20° B 9.04, CD 7.93 ± .30 (247).
Aluminum chloride·o-chloronitrobenzene [AlCl$_3$·C$_6$H$_4$ClNO$_2$], B 9.48 (700).
Aluminum chloride·p-chloronitrobenzene [AlCl$_3$·C$_6$H$_4$ClNO$_2$], B 7.79 (700).
Aluminum chloride·ethylamine [AlCl$_3$·C$_2$H$_5$NH$_2$], B 6.86 (194); B 6.82 (195);
 B 20° 6.94 (247).
Aluminum chloride·ethyl ether [AlCl$_3$·(C$_2$H$_5$)$_2$O], B 6.54 (194); B 6.45 (195);
 B 25° 6.68, CD 20° 5.94 (247).
Aluminum chloride·nitrobenzene [AlCl$_3$·C$_6$H$_5$NO$_2$], B 9.05 (194); B 8.99
 (195); B 20° 9.25 (247).
Aluminum chloride·o-nitrotoluene [AlCl$_3$·CH$_3$C$_6$H$_4$NO$_2$], S 8.92 (700).
Aluminum chloride·p-nitrotoluene [AlCl$_3$·CH$_3$C$_6$H$_4$NO$_2$], S 9.68 (700).
Aluminum laurate [(C$_{11}$H$_{23}$CO$_2$-)$_3$Al], B 25° 4.06 (371).
Aluminum myristate [(C$_{13}$H$_{27}$CO$_2$-)$_3$Al], B 25° 4.42 (371).
Aluminum oleate [(C$_{17}$H$_{33}$CO$_2$-)$_3$Al], B 25° 4.29 (371).
Aluminum palmitate [(C$_{15}$H$_{31}$CO$_2$-)$_3$Al], B 25° 5.22 (371).
Aluminum stearate [(C$_{17}$H$_{35}$CO$_2$-)$_3$Al], B 25° 5.49 (371).
Aluminum n-valerate [(C$_4$H$_9$CO$_2$-)$_3$Al], B 25° 3.44 (371).

Antimony
 Triphenylstibine [(C$_6$H$_5$)$_3$Sb], B 14° 0.57 (214).
 Triphenylstibine dichloride [(C$_6$H$_5$)$_3$SbCl$_2$], B 25° 1.19 (769); B 25° 0 (785).
 Triphenylstibine dihydroxide [(C$_6$H$_5$)$_3$Sb(OH)$_2$], B 25° 0 (785).
 Triphenylstibine hydroxychloride [(C$_6$H$_5$)$_3$SbOHCl], D 40° 2.96 ± .05 (786).
 Triphenylstibine sulfide [(C$_6$H$_5$)$_3$SbS], B 25° 5.40 (786).
 Tri-p-tolylstibine oxide [(CH$_3$C$_6$H$_4$-)$_3$SbO], B 25° 2.0 ± .1, D 40° 2.3 ±.1
 (786).

Arsenic
 Arsenic N,N-diethyldithiocarbamate [As(S$_2$CN(C$_2$H$_5$)$_2$)$_3$], B or CD 20°
 4.5$_8$ (656).
 Arsenic N,N-dimethyldithiocarbamate [As(S$_2$CN(CH$_3$)$_2$)$_3$], B 39° 4.7$_1$ ± .08
 (656).
 Arsenic N,N-di-n-propyldithiocarbamate [As(S$_2$CN(C$_3$H$_7$)$_2$)$_3$], B 20° 5.0$_2$
 (656).
 Arsenic O-ethylxanthate [·As(S$_2$COC$_2$H$_5$)$_3$], B or CD 20° 1.5$_2$ (656).
 Arsenic O-isopropylxanthate [As(S$_2$COC$_3$H$_7$)$_3$], B 20° 1.5$_7$ (656).
 Arsenobenzene [C$_6$H$_5$As=AsC$_6$H$_5$], B 25° ca. 0 (650).
 β-Chlorovinyldichloroarsine [ClCH=CHAsCl$_2$], B 20° 1.77 (612).
 β,β'-Dichlorodivinylchloroarsine [(ClCH=CH-)$_2$AsCl], B 20° 1.45 (612).
 Diphenylchloroarsine [(C$_6$H$_5$)$_2$AsCl], Hx 20° 2.70 (610).
 Diphenylcyanoarsine [(C$_6$H$_5$)$_2$AsCN], Hx 20° 4.19 (610).

5

Ethyldichloroarsine $[C_2H_5AsCl_2]$, B 20° 2.51 (612).
Phenarsazine chloride $[HN(C_6H_4)_2AsCl]$, B 20° 2.26 (610).
β, β', β''-Trichlorotrivinylarsine $[(ClCH=CH)_3As]$, B 20° 0.39 (612).
Triphenylarsine $[(C_6H_5)_3As]$, B 18° 1.07 (214).
Triphenylarsine dichloride $[(C_6H_5)_3AsCl_2]$, B 25° 0 (785).
Triphenylarsine hydroxychloride $[(C_6H_5)_3AsOHCl]$, D 40° 9.2 (786).
Triphenylarsine oxide $[(C_6H_5)_3AsO]$, B 25° 5.50 (786).
Triphenylarsine oxide hydrate $[(C_6H_5)_3AsO·H_2O]$, B 25° 5.75 (786).

Beryllium

Beryllium acetate, basic $[(CH_3CO_2-)_6Be_4O]$, B 25 and 45° 0 (257).
Beryllium acetoacetate $[(CH_3COCH_2CO_2-)_2Be]$, B 3.07 (746).
Beryllium acetylacetonate $[(C_5H_7O_2-)_2Be]$, CT 20-45° 0 (257).
Beryllium bromide·ethyl ether $[BeBr_2·2(C_2H_5)_2O]$, B 20° 7.57 (194, 247);
    B 7.52 (195).
Beryllium chloride·ethyl ether $[BeCl_2·2(C_2H_5)_2O)]$, B 6.84 (194); B 6.74
    (195); B 20° 6.71 (247).

Bismuth

Triphenylbismuth $[(C_6H_5)_3Bi]$, B 16° 0 (214).
Triphenylbismuth dichloride $[(C_6H_5)_3BiCl_2]$, B 25° 1.17 (769); B 25° 0
    (785).
Triphenylbismuth dinitrate $[(C_6H_5)_3Bi(NO_3)_2]$, B 25° 3.26 ± .15 (785).

Boron

n-Amyl borate $[(C_5H_{11})_3BO_3]$, 25° B 0.79, D 0.76 (438).
n-Amylboric acid $[(C_5H_{11}B(OH)_2]$, 25° B 1.90, D 1.92 (438).
Boron trichloride·acetonitrile $[BCl_3·CH_3CN]$, B 7.65 (194); B 7.62 (195);
    B 20° 7.74 ± .16 (247).
Boron trichloride·ethyl ether $[BCl_3·2(C_2H_5)_2O]$, B 5.98 (194); B 5.96 (195);
    B 20° 6.04 (247).
Boron trichloride·ethyl sulfide $[BCl_3·(C_2H_5)_2S]$, B 25° 6.00 ± .01 (817).
Boron trichloride·propionitrile $[BCl_3·C_2H_5CN]$, B 7.75 (194); B 7.64 (195);
    B 20° 7.83 (247).
Boron trichloride·trimethylamine $[BCl_3·(CH_3)_3N]$, B 25° 6.23 ± .02 (817).
Boron trichloride·trimethylphosphine $[BCl_3·(CH_3)_3P]$, B 25° 7.03 ± .01
    (817).
Boron trichloride·triphenylphosphine $[BCl_3·(C_6H_5)_3P]$, B 25° 7.01 ± .06
    (817).
Boron trifluoride·ethyl ether $[(BF_3·(C_2H_5)_2O]$, B 25° 4.92 ± .38 (789); B
    25° 5.29 ± .03 (817).
Boron trifluoride·methyl ether $[BF_3·(CH_3)_2O]$, B 25° 4.35 ± .12 (789).
Boron trifluoride·methyl ethyl ether $[BF_3·CH_3OC_2H_5]$, B 25° 5.07 ± .08
    (789).
Boron trifluoride·trimethylamine $[BF_3·(CH_3)_3N]$, B 25° 5.76 ± .02 (817).
n-Butyl borate $[(C_4H_9)_3BO_3]$, 25° B 0.77, D 0.79 (438).
sec-Butyl borate $[(C_4H_9)_3BO_3]$, B 25° 0.85 (687).
n-Butylboric acid $[C_4H_9B(OH)_2]$, 25° B 1.85, D 1.94 (438).
Isobutyl borate $[(C_4H_9)_3BO_3]$, B 25° 0.85 (687).
Phenylboric acid $[C_6H_5B(OH)_2]$, 25° B 1.64, D 1.72 (438).
n-Propyl borate $[(C_3H_7)_3BO_3]$, B 25° 0.77 (687).
Tetra-n-butylammonium-hydroxide·triphenylboron $[(C_4H_9)_4NOH·B(C_6H_5)_3]$,
    B 25° 19.7 (472).
Triethyl borate $[(C_2H_5)_3BO_3]$, B 20° 0.75 (412).

Triisoamyl borate $[(C_5H_{11})_3BO_3]$, B 20° 0.81 (412).

Cadmium

Cadmium N,N-diisobutyldithiocarbamate $[Cd(S_2CN(C_4H_9)_2)_2]$, B 21° 1.6$_8$ (690).

Cadmium oleate $[(C_{17}H_{33}CO_2-)_2Cd]$, B 25° 4.37 (371).

Cadmium stearate $[(C_{17}H_{35}CO_2-)_2Cd]$, B 25° 4.80 (371).

Chromium

Chromic N,N-di-n-butyldithiodicarbamate $[Cr(S_2CN(C_4H_9)_2)_3]$, B 30° 1.9$_8$ (690).

Chromic dinitrosyl diethyldithiocarbamate $[(NO)_2Cr(S_2CN(C_2H_5)_2)_2]$, B 23° 5.1$_9$ (691).

Cobalt

Cobaltic N,N-di-n-propyldithiocarbamate $[Co(S_2CN(C_3H_7)_2)_3]$, B 20° 1.6$_1$ (690).

Cobaltic N,N-isobutyldithiocarbamate $[(Co(S_2CN(C_4H_9)_2)_3]$, B 23° 1.9$_2$ (690).

Cobaltic O-isopropylxanthate $[Co(S_2COC_3H_7)_3]$, B 25° 2.9$_2$ (690).

Cobalti-α-nitroso-β-naphthol $[(C_{10}H_6(NO)O-)_3Co]$, B or Hx 3.8 (237).

Cobaltous chloride·triethylphosphine $[CoCl_2·2(C_2H_5)_3P]$, B 20° 8.7 (483).

Copper

Cupric N,N-di-n-propyldithiocarbamate $[Cu(S_2CN(C_3H_7)_2)_2]$, B 30° 1.6$_7$ (690).

Gallium

Gallium trichloride·benzonitrile $[GaCl_3·C_6H_5CN]$, B R.T. 8.65 (744).

Gallium trichloride·benzoyl chloride $[GaCl_3·C_6H_5COCl]$, CT R.T. 6.85 (744).

Gallium trichloride·p-nitrotoluene $[GaCl_3·CH_3C_6H_4NO_2]$, B R.T. 9.16 (744).

Germanium

Triphenylgermanium bromide $[(C_6H_5)_3GeBr]$, S 2.5 (740); B 25° 2.35 (769).

Gold

Diethylgold bromide $[(C_2H_5)_2AuBr]$, B 25° >1 (408); B 25° ca. 0 (415); CT 25° 1.32 (525).

Di-n-propylgold cyanide $[(C_3H_7)_2AuCN]$, CT 25° 1.47 (525).

n-Propylgold dibromide $[C_3H_7AuBr_2]$, CT 25° 5 to 6 (408); CT 25° ca. 6.0 (415).

Iron

Ferric N,N-di-n-propyldithiocarbamate $[Fe(S_2CN(C_3H_7)_2)_3]$, B 20° 1.6$_1$ (690).

Ferric O-ethylxanthate $[Fe(S_2COC_2H_5)_3]$, B 25° 2.7$_8$ (690).

Ferric O-isopropylxanthate $[Fe(S_2COC_3H_7)_3]$, B 20° 2.8$_8$ (690).

Ferric α-nitroso-β-naphthol $[Fe(C_{10}H_6(NO)O-)_3]$, B or Hp **3.1** (237).
Ferric nitrosyl diisobutyldithiocarbamate $[(NO)Fe(S_2CN(C_4H_9)_2)_2]$, B 23° **4.2$_6$** (691).
Iron pentacarbonyl $[Fe(CO)_5]$, B 12° **0.64** (152, 153, 154); B 20° **0.81** (228).

Lead

Diethyllead dichloride $[(C_2H_5)_2PbCl_2]$, B 25° **4.70** (686, 739).
Ethyl-triphenyllead $[(C_6H_5)_3PbC_2H_5]$, B 22° **0.81** (792).
Phenyl-triethyllead $[(C_2H_5)_3PbC_6H_5]$, B 22° **0.86** (792).
Triethyllead bromide $[(C_2H_5)_3PbBr]$, B 25° **4.46** (686, 740); T 23° **4.88** (792).
Triethyllead chloride $[(C_2H_5)_3PbCl]$, B 25° **4.39** (686, 739); CT 22° **4.66** (792).
Trimethyllead chloride $[(CH_3)_3PbCl]$, B 25° **4.47** (686, 740).
Triphenyllead $[((C_6H_5)_3Pb)_2]$, B 25° **0** (686).
Triphenyllead bromide $[(C_6H_5)_3PbBr]$, B 25° **4.21** (686, 740).
Triphenyllead chloride $[(C_6H_5)_3PbCl]$, B 25° **4.21** (686, 739); B 22° **4.32** (792).
Triphenyllead hydroxide $[(C_6H_5)_3PbOH]$, B 22° ca. **2.4** (792).
Triphenyllead iodide $[(C_6H_5)_3PbI]$, B 25° **3.73** (686, 740).

Lithium

n-Butyllithium $[C_4H_9Li]$, B **0.97** (842).

Magnesium

Magnesium oleate $[(C_{17}H_{33}CO_2-)_2Mg]$, B 25° **2.96** (371).
Magnesium palmitate $[(C_{15}H_{31}CO_2-)_2Mg]$, B 25° **3.32** (371).
Magnesium stearate $[(C_{17}H_{35}CO_2-)_2Mg]$, B 25° **3.67** (371).

Mercury

n-Amylmercuric bromide $[C_5H_{11}HgBr]$, 25° B **3.47**, D **3.53** (754).
Benzylmercuric chloride $[C_6H_5CH_2HgCl]$, S **3.0** (740); B 25° **3.05** (769).
n-Butylmercuric bromide $[C_4H_9HgBr]$, D 25° **3.45** (754).
p-Chlorophenylmercuric bromide $[ClC_6H_4HgBr]$, D 25° **1.57** (754).
Di-p-bromophenylmercury $[(BrC_6H_4)_2Hg]$, Dec 142° **0.92** (352).
Di-p-chlorophenylmercury $[(ClC_6H_4)_2Hg]$, Dec 142° **1.15** (352).
Diethylmercury $[(C_2H_5)_2Hg]$, B 15° **0.39** (214).
Di-p-fluorophenylmercury $[(FC_6H_4)_2Hg]$, Dec 142° **0.87** (352).
Diphenylmercury $[(C_6H_5)_2Hg]$, B 14° **0** (214); B 25° **0.41**, Dec 25° **0.44**, 142° **0.54** (352); D 25° **0.42** (413).
Di-p-tolylmercury $[(CH_3C_6H_4)_2Hg]$, Dec 142° **0.74** (352).
Mercurous trichloroacetate $[(Cl_3CCO_2-)_2Hg_2]$, B 25° **1.50** ± .20 (755).
Phenylmercuric bromide $[C_6H_5HgBr]$, D 50° **3.06** (754).
p-Tolylmercuric bromide $[CH_3C_6H_4HgBr]$, D 25 and 50° **3.39** (754).

Nickel

Bis-diethyl-phenylphosphinenickel dichloride $[((C_2H_5)_2(C_6H_5)P)_2NiCl_2]$, B 20° ca. **0** (482).
Bis-triethylphosphinenickel dibromide $[((C_2H_5)_3P)_2NiBr_2]$, B 20° ca. **0** (482).
Bis-triethylphosphinenickel dinitrate $[((C_2H_5)_3P)_2Ni(NO_3)_2]$, B 20° **8.85** (482).
Nickel carbonyl $[Ni(CO)_4]$, CT 0° **0** (261); CT 0° ca. **0.3** (322).
Nickel p-chlorophenyl-n-butylglyoxime, acid $[(C_{12}H_{14}ClN_2O_2)_2Ni]$, α- , B 25° **1.8** (409).
Nickel methyl-benzylglyoxime, acid $[(C_{10}H_{11}N_2O_2)_2Ni]$, α- , B 25° **1.3** (409); β-, B 25° **1.6** (409).

Nickel methyl-n-butylglyoxime, acid [$(C_7H_{13}N_2O_2)_2Ni$], α-, B 25° 1.3
(409); β-. B 25° 1.3 (409).

Nickel methyl-n-propylglyoxime, acid [$(C_6H_{11}N_2O_2)_2Ni$], α- B 25° 1.4
(409); β- , B 25° 1.5 (409).

Nickelous O,O-diethyldithiophosphate [$(C_2H_5O)_2PS_2)_2Ni$], B 31° 2.2 (837).

Nickelous diethyldithiophosphinate [$(C_2H_5)_2PS_2)_2Ni$] α- , B 31° 1.66
(837); β-, CT 21.5° 1.70 (837).

Nickelous N,N-diisobutyldithiocarbamate [$Ni(S_2CN(C_4H_9)_2)_2$], B 30° 1.7₈
(690).

Nickelous diisopropyldithiophosphinate [$(C_3H_7)_2PS_2)_2Ni$], B 35.3° 1.67
(837).

Nickelous di-n-propyl dithiophosphate [$(C_3H_7O)_2PS_2)_2Ni$], B 31° 2.0 (837).

Nickelous O-isobutylxanthate [$Ni(S_2COC_4H_9)_2$], B 22° 2.3₃ (690).

Nickelous N,N-isohexydithiocarbamate [$Ni(S_2CNHC_6H_{13})_2$], B 23° 1.9₁
(690).

Nickelous O-isopropylxanthate [$Ni(S_2COC_3H_7)_2$], B 30° 2.6₀ (690).

Tri-n-propylphosphinenickel dichloride [$((C_3H_7)_3P)_2 NiCl_2$], B 20° ca. 0
(482).

Palladium

Bis-diethylsulfide palladium dichloride [$((C_2H_5)_2S)_2PdCl_2$], α- , B 20° 2.19
(422).

Platinum

Bis-dibenzylsulfide platinum dibromide [$((C_6H_5CH_2)_2S)_2PtBr_2$], β-, B 20°·
8.7 (422).

Bis-dibenzylsulfide platinum dichloride [$((C_6H_5CH_2)_2S)_2PtCl_2$], β-, B 20°
7.8 (422).

Bis-dibenzylsulfide platinum diiodide [$((C_6H_5CH_2)_2S)_2PtI_2$], α- , B 20°
2.37 (422).

Bis-di-n-butylsulfide platinum dibromide [$((C_4H_9)_2S)_2PtBr_2$], α-, B 20°
2.19 (422).

Bis-di-n-butylsulfide platinum dichloride [$((C_4H_9)_2S)_2PtCl_2$], α-, B 20°
2.35 (422); β- , B 20° 9.2 (422).

Bis-di-sec-butylsulfide platinum dichloride [$((C_4H_9)_2S)_2PtCl_2$], α-, B 20°
2.39 (422).

Bis-di-n-butylsulfide platinum diiodide [$((C_4H_9)_2S)_2PtI_2$], α-, B 20° 2.18
(422).

Bis-diethylphenylphosphine platinum dichloride [$((C_2H_5)_2(C_6H_5)P)_2PtCl_2$],
α-, B 20° ca. 0 (481).

Bis-diethylselenide platinum dichloride [$((C_2H_5)_2Se)_2PtCl_2$], α- , B 20°
2.41 (422); β- , B 20° 9.1 (422).

Bis-diethylsulfide platinum dibromide [$((C_2H_5)_2S)_2PtBr_2$], α- , B 20° 2.26
(422); β-, B 20° 8.9 (422).

Bis-diethylsulfide platinum dichloride [$((C_2H_5)_2S)_2PtCl_2$], α-, B 20° 2.41
(422); β- , B 20° 9.5 (422).

Bis-diethylsulfide platinum diiodide [$((C_2H_5)_2S)_2PtI_2$], α- , B 20° 2.41 (422).

Bis-diethylsulfide platinum dinitrite [$((C_2H_5)_2S)_2Pt(NO_2)_2$], β- , B 20°
13.7 (422).

Bis-diisobutylsulfide platinum dichloride [$((C_4H_9)_2S)_2PtCl_2$], α- , B 20°
2.41 (422); β-, B 20° 9.0 (422).

Bis-diisobutylsulfide platinum dinitrite [$((C_4H_9)_2S)_2Pt(NO_2)_2$], α- , B 20°
2.51 (422); β-, B 20° 13.7 (422).

Bis-diisopropylsulfide platinum dichloride [$((C_3H_7)_2S)_2PtCl_2$], α-, B 20°
2.39 (422).

9

Bis-diphenyltelluride platinum dichloride $[((C_6H_5)_2Te)_2PtCl_2]$, cis-, B
   $20°$ $6.0 ± .5$ (544).
Bis-diphenylsulfide platinum dichloride $[((C_6H_5)_2S)_2PtCl_2]$, cis-, B $20°$
   5.5 (544).
Bis-di-n-propylsulfide platinum dibromide $[((C_3H_7)_2S)_2PtBr_2]$, α-, B $20°$
   2.32 (422).
Bis-di-n-propylsulfide platinum dichloride $[((C_3H_7)_2S)_2PtCl_2]$, α-, B $20°$
   2.39 (422).
Bis-di-n-propylsulfide platinum dinitrate $[((C_3H_7)_2S)_2Pt(NO_3)_2]$, α-, B $20°$
   3.93 (422); β-, B $20°$ 11.9 (422).
Bis-di-n-propylsulfide platinum dinitrite $[((C_3H_7)_2S)_2Pt(NO_2)_2]$, α-, B $20°$
   2.48 (422); β-, B $20°$ 13.1 (422).
Bis-tri-n-butylarsine platinum dichloride $[((C_4H_9)_3As)_2PtCl_2]$, α-, B $20°$
   ca. 0 (481).
Bis-tri-n-butylphosphine platinum dichloride $[((C_4H_9)_3P)_2PtCl_2]$, α-, B
   $20°$ ca. 0 (481); β-, B $20°$ 11.5 (481).
Bis-tri-n-butylstibine platinum dichloride $[((C_4H_9)_3Sb)_2PtCl_2]$, β-, B,$20°$
   10.9 (481).
Bis-triethylarsine platinum dichloride $[((C_2H_5)_3As)_2PtCl_2]$, α-, B $20°$ ca.
   0 (481); β-, B $20°$ 10.5 (481).
Bis-triethylphosphine platinum dibromide $[((C_2H_5)_3P)_2PtBr_2]$, α-, B $20°$
   ca. 0 (481); β-, B $20°$ 11.2 (481).
Bis-triethylphosphine platinum dichloride $[((C_2H_5)_3P)_2PtCl_2]$, α-, B $20°$
   ca. 0 (481); β-, B $20°$ 10.7 (481).
Bis-triethylphosphine platinum diiodide $[((C_2H_5)_3P)_2PtI_2]$, α-, B $20°$ ca.
   0 (481); β-, B $20°$ 8.2 (481).
Bis-triethylphosphine platinum dinitrate $[((C_2H_5)_3P)_2Pt(NO_3)_2]$, α-, B $20°$
   ca. 0 (481).
Bis-triethylphosphine platinum dinitrite $[((C_2H_5)_3P)_2Pt(NO_2)_2]$, α-, B $20°$
   ca. 0 (481).
Bis-triethylstibine platinum dichloride $[((C_2H_5)_3Sb)_2PtCl_2]$, β-, B $20°$
   9.2 (481).
Bis-triethylstibine platinum diiodide $[((C_2H_5)_3Sb)_2PtI_2]$, α-, B $20°$ ca. 0
   (481).
Bis-triethylstibine platinum dinitrite $[((C_2H_5)_3Sb)_2Pt(NO_2)_2]$, α-, B $20°$
   ca. 0 (481).
Bis-triphenylstibine platinum dichloride $[((C_6H_5)_3Sb)_2PtCl_2]$, β-, B $20°$
   9.3 (481).
Bis-tri-n-propylphosphine platinum dichloride $[((C_3H_7)_3P)_2PtCl_2]$, α-,
   B $20°$ ca. 0 (481); β-, B $20°$ 11.5 (481).
Trimethyl platinum bromide $[(CH_3)_3PtBr]$, ca. 1 (861).
Trimethyl platinum chloride $[(CH_3)_3PtCl]$, ca. 1 (861).
Trimethyl platinum iodide $[(CH_3)_3PtI]$, ca. 1 (861).

Silicon

Polymethylpolysiloxanes, linear $[(CH_3)_3Si(OSi(CH_3)_2)_nCH_3]$, Lqd. $25°$
   (n = 1) 0.43, (n = 2) 0.799, (n = 3) 0.43, (n = 4) 0.932, (n = 5) 1.167,
   (n = 6) 1.143, (n = 7) 1.106 (823); Lqd. $20°$ (n = 1) 0.74, (n = 2) 0.99,
   (n = 3) 1.22, (n = 4) 1.40, (n = 5) 1.58 (843); cyclic $[((CH_3)_2SiO)_n]$,
   Lqd. $20°$ (n = 4) 1.09, (n = 5) 1.35, (n = 6) 1.56, (n = 7) 1.78, (n = 8)
   1.96 (843).
Tetraethylsilane $[(C_2H_5)_4Si]$, CT $20°$ 0 (766).
Tetramethylsilane $[(CH_3)_4Si]$, Lqd. $25°$ 0 (823).
Triethyl-isoamylsilane $[C_5H_{11}Si(C_2H_5)_3]$, CT $27°$ 0 (766).
Triethylphenylsilane $[C_6H_5Si(C_2H_5)_3]$, CT $27°$ 0.7 (766).
Triethylsilanol $[(C_2H_5)_3SiOH]$, CT $22°$ 1.50 (790).

Triethylsilyl chloride [(C$_2$H$_5$)$_3$SiCl], CT 20° 2.07 (790).
Triethylsilyl ether [((C$_2$H$_5$)$_3$Si)$_2$O], CT 24.2° 0.63 (790).
Triethylsilane [(C$_2$H$_5$)$_3$SiH], CT 25° ca. 0 (766).
Triphenylsilane [(C$_6$H$_5$)$_3$SiH], CT 25° 1 (766).
Triphenylsilanol [(C$_6$H$_5$)$_3$SiOH], CT 22.6° 1.45 (790).

Tellurium
  Bis-di-p-tolyltellurium chloride oxide [((CH$_3$C$_6$H$_4$)$_2$TeCl)$_2$O], D 25° 6.1±
    .2 (784).
  Dimethyltellurium diiodide [(CH$_3$)$_2$TeI$_2$], B 25° 2.26 (574).
  Di-p-tolyltellurium dibromide [(CH$_3$C$_6$H$_4$)$_2$TeBr$_2$], B 25° 3.21 (784).
  Di-p-tolyltellurium dichloride [(CH$_3$C$_6$H$_4$)$_2$TeCl$_2$], B 25° 2.98 (784).
  Di-p-tolyltellurium oxide [(CH$_3$C$_6$H$_4$)$_2$TeO], D 40° 3.93 (786).
  Phenyl telluride [(C$_6$H$_5$)$_2$Te], B 21° 1.13$_5$ (103).

Tin
  Ethyl-triphenyltin [(C$_6$H$_5$)$_3$SnC$_2$H$_5$], B 22° 0.73 (791).
  Phenyl-triethyltin [(C$_2$H$_5$)$_3$SnC$_6$H$_5$], CT 22° 0.5 (791).
  Stannic bromide·triphenylmethyl bromide [SnBr$_4$·(C$_6$H$_5$)$_3$CBr], B 25° 2.7
    (810).
  Stannic chloride·acetone [SnCl$_4$·2(CH$_3$)$_2$CO], B 15° 7.7 (262).
  Stannic chloride·acetophenone [SnCl$_4$·2C$_6$H$_5$COCH$_3$], B 18° 8.7 (262).
  Stannic chloride·benzaldehyde [SnCl$_4$·2C$_6$H$_5$CHO], B 18° 7.5 to 8.1 (262).
  Stannic chloride·benzonitrile [SnCl$_4$·2C$_6$H$_5$CN], B 15° 6.81 (262).
  Stannic chloride·ethyl acetate [SnCl$_4$·2CH$_3$COOC$_2$H$_5$], B 18° 7.3 (799).
  Stannic chloride·ethyl ether [SnCl$_4$·2(C$_2$H$_5$)$_2$O], B 15° 3.60 (262).
  Tetraethyltin [(C$_2$H$_5$)$_4$Sn], CT 25° 0 (321); B 22° ca. 0 (791).
  Triethyltin bromide [(C$_2$H$_5$)$_3$SnBr], CT 25° 3.32 (321).
  Triethyltin chloride [(C$_2$H$_5$)$_3$SnCl], CT 25° 3.44 (321).
  Triethyltin hydroxide [(C$_2$H$_5$)$_3$SnOH], B 22° 1.91 (791).
  Triphenyltin bromide [(C$_6$H$_5$)$_3$SnBr], B 25° 3.15 (791).
  Triphenyltin chloride [(C$_6$H$_5$)$_3$SnCl], B 25° 3.28 (769).
  Triphenyltin hydroxide [(C$_6$H$_5$)$_3$SnOH], B 22° 1.98 (791).

Titanium
  Titanium tetrachloride·benzonitrile [TiCl$_4$·C$_6$H$_5$CN], B 15° 6.16 (262).
  Titanium tetrachloride·propionitrile [TiCl$_4$·C$_2$H$_5$CN], B 16° 6.05 (262).

Zinc
  Zinc O-n-butylxanthate [Zn(S$_2$COC$_4$H$_9$)$_2$], B 30° 2.5$_6$ (690).
  Zinc N,N-di-n-propyldithiocarbamate [Zn(S$_2$CN(C$_3$H$_7$)$_2$)$_2$], B 22° 1.4$_8$
    (690).
  Zinc O-isoamylxanthate [Zn(S$_2$COC$_5$H$_{11}$)$_2$], B 30° 2.5$_1$ (690).
  Zinc N,N-isohexyldithiocarbamate [Zn(S$_2$CNHC$_6$H$_{13}$)$_2$], B 23° 1.9$_1$ (690).

The empirical formulas are arranged in numerical sequence, carbon being given first, hydrogen, if present, second, and the other constituents in alphabetical order.

$CBrCl_3$	Trichlorobromomethane, B 0 (208).
CBrN	Cyanogen bromide, B 20° 2.94 ± .02 (794).
CBr_2Cl_2	Dichlorodibromomethane, Hp or B 25° 0 (546).
CBr_3F	Fluorotribromomethane, B 0 (208).
CCl_2F_2	Difluorodichloromethane, 305–470°K, 0.51 (319).
CCl_2O	Phosgene (carbonyl chloride), 303–425°K 1.18 (387); CT 0° 1.099 (427).
CCl_2S	Thiophosgene, 303–414°K 0.28 ± .02 (636).
CCl_3F	Fluorotrichloromethane, 299–376°K 0.45 (319).
CCl_3NO_2	Chloropicrin, B or CT 25° 1.8 (274); 303–425°K 1.18 (387); 25–50° Hp 1.79, B 1.80 (451); Hx 1.91 (611).
CCl_4	Carbon tetrachloride, Lqd. 0 (8); B < 0.1 (23); 361–432°K 0 (20); B 25° 0 (31); S 0 (349).
CF_4	**Carbon tetrafluoride**, 193–368°K 0 (514).
$CHBr_3$	**Bromoform**, B 25° 0.9₀ (127); B or CT 25° 0.99 (274); Lqd. 0.91, B 1.06, Cf 0.85, CB 0.91, NB 1.44 (493).
$CHCl_2F$	**Fluorodichloromethane**, 305–424°K 1.29 (319).
$CHCl_3$	**Chloroform**, Lqd. 1.256 (8); 343–443°K 0.9₅ (20); CT 25° 1.15 (24); B 25° 1.10 (31); S 1.14 (39); 297–411°K 1.05 (50); Lqd. and Hx-90 to 60° 1.05 (51); B 18° 1.27, E 1.55 (116); B 19° 1.55 (121); B 1.27, E 1.57 (122); B 25° 1.1₈ (127); B or CT 25° 1.10 (274); Lqd. or Hx -60° to -20° 1.20 (348, 349); 298°K 1.06, Lqd. 1.10, B 1.22, T 1.24, CB 1.18, NB 1.48 (493); CT -18 to 46° 1.21 (504); Vap. 1.86 (735); gasoline 15° 1.16 (854).
CHF_3	**Fluoroform**, 298–368°K 1.59 (445).
CHI_3	**Iodoform**, 25° Hx 1.0₀, B 0.9₅ (127); B or CT 25° 0.99 (274).
CHN	**Hydrogen cyanide**, B 20° 2.65 (97); 303–474°K 2.1 (113); 292–424°K 2.88 (158); 16–48° B 2.54, pX 2.60 (171); 301–470°K 2.93₂ (387); 298–368°K 3.03 (513).
CHN_3O_6	**Nitroform**, CT 25° 2.71 (652).
CH_2Br_2	**Methylene bromide**, 322–420°K 1.914 (46); B 25° 1.89 (81); 25° Hx 1.4₂, B 1.3₈ (127); B or CT 25° 1.40 (274); Hx 0–40° 1.28, E 20° 1.68 (546); Vap. 2.27 (735).
CH_2ClNO_2	**Chloronitromethane**, 412–484°K 2.91 (758).
CH_2Cl_2	**Methylene chloride**, 332–414°K 1.5₉ (20); 303–426°K 1.62₁ (46); B 25° 1.61 (81); CT -89 to 42° 1.48 (126); B 25° 1.5₅ (127); B or CT 25° 1.57 (274); -10 to 30° Lqd. or CT 1.78 (349); 0° B 1.50, CD 1.13 (368); CT -18 to 46° 1.53 to 1.59 (504); Vap. 2.39 (735).
CH_2I_2	**Methylene iodide**, B 25° 2.12 (81); 25° Hx 1.1₄, B 1.1₀ (127); B or CT 25° 1.10 (274).
CH_2N_2	Cyanamide, B 20° 3.8 (284).
CH_2N_4	1, 2, 3, 4-Tetrazole, D 25° 5.11 (787).
CH_2O	Formaldehyde, 420–520°K 2.27 (783).
CH_2O_2	Formic acid, B 22° 1.45 (105); B 25° 1.19 ± .02 (147); 345–423°K 1.51 (206); D 25° 2.07 (399); B 30° 1.77 (699); D 1.98 (866).
CH_3Br	Methyl bromide, 301–417°K 1.786 (124); Hx -102 to 38° 1.45 (126); B or CT 25° 1.82 (274); 306–406°K 1.78 (388); 303–417°K

	1.75 (429); 291-416°K 1.79 ± .01 (536).
CH_3Cl	Methyl chloride, 291-415°K 1.9$_7$ (20); 293°K Kerr 1.66 (25); 292-474°K 1.69 (50); 298-418°K 1.85 ± .01 (87); 290-456°K 1.86 (114); 126-404°K 2.002 (124); Hx -93 to 42° 1.56, CT -46 to 60° 1.65 (126); Gas 1.861 ± .008 (133); 298 to 418°K 1.86 ± .005 (254); B or CT 25° 1.86 (274); -60 to -20° Lqd. or Hx 1.85 (348, 349); 307-414°K 1.83 (429).
CH_3ClO_2S	Methyl chlorosulfite, B 25° 2.30, 45° 2.33 (455).
CH_3F	Methyl fluoride, 224-498°K 1.808 (388).
CH_3I	Methyl iodide, 304-395°K 1.313 (45); B 25° 1.6 (62); 304-417°K 1.623 (124); Hx -90 to 34° 1.35 (126); B or CT 25° 1.66 (274); 305-494°K 1.59 (388); 304-416°K 1.60 (429); B 5-28° 1.35 to 1.37 (504); B 20° 1.41 (528); 295-337°K 1.64 ± .01 (536); CT 20° 1.56 (711, 750).
CH_3NO	Formamide, 425-449°K 3.2$_2$ (270).
CH_3NO_2	Nitromethane, B 3.05 (79); B 25° 3.13 (265); B or CT 25° 3.20 (274); 25° B 3.10, CT 3.10 (334, 336, 390); 339-494°K 3.42 (387); Hp 25-50° 3.18 (451); 337-454°K 3.54 ± .01 (536); Vap. 3.50 (739).
CH_3NO_3	Methyl nitrate, B 20° 2.85 (279).
CH_3N_5	5-Amino-1, 2, 3, 4-tetrazole, D 25° 5.71 (787).
CH_4	Methane, 292-415°K 0 (20); 298°K 0 (394).
CH_4N_2O	Urea, D 47° ca. 8.6 (578); D 25° 4.56 (763).
CH_4N_2S	Thiourea, D 47° ca. 7.6 (578); D 25° 4.89 (763).
CH_4O	Methyl alcohol, 351-451°K 1.61 (1); B 1.66 (54); B 10-70° 1.64 (55); 345-502°K 1.680 (83); B 1.6 (140); B 22° 1.66$_4$ ± .02 (203); 308-482°K 1.69 (424); B 1.66 (564); B 1.69 (655); B 30° 1.62 (734); Lqd. 30-40° 3.01 (816).
CH_5N	Methyl amine, 338-458°K 1.23 ± .02 (139, 187, 254); 296-362°K 0.99 (165); 288-417°K 1.32 (537); 298°K 1.28, B 25° 1.46, Lqd. -10 to +25° 1.08 (852).
CH_6N_2	Methylhydrazine, B 15° 1.68 ± .14 (457).
ClN	Cyanogen iodide, B 20° 3.71 ± .02 (776).
CN_4O_8	Tetranitromethane, B 25° < 0.2 (61); CT 25° 0.19 (265); CT 25° 0.71 (652).
CO	Carbon monoxide, 84°K 0.128 ± .007 (2); Gas 0.1180 ± .0016 (5); 288-381°K 0.124 ± .01 (22); 292-572°K 0.119 ± .029 (43); 90-391°K 0.10 (67); 298°K 0.10 (394)
COS	Carbon oxysulfide (carbonyl sulfide), 202-365°K 0.650 (67), Stark 0.72 (825); Stark, $C^{12}OS$ 0.732 ± .007, $C^{13}OS$ 0.722 ± .007 (869); 265-333°K 0.720 ± .005 (851).
COSe	Carbon oxyselenide, Stark 0.752 ± .007 (871).
CO_2	Carbon dioxide, 296-426°K 0.303 ± .145 (1); 200°K 0.132 (2); Gas 0.1420 ± .0017 (5); 293°K 0.20 ± .05 (18); 292-383°K 0.145 ± .03 (22); 294-477°K 0.208 ± .029 (43); 259-456°K 0 (56); 293-415°K 0 (158); 273-373°K 0 (325); 298°K 0 (394).
CS_2	Carbon disulfide, Lqd. 0 (8); 25° B 0.06, Hx 0.08 (63); 302-490°K 0.326 (67); 305-413°K 0 (75); 302-424°K 0.020 (135); 322-489°K 0 (149); B, Hp or CT 20° 0 (220); 25° B ca. 0, CD 0, CB 0.49, NB 1.20 (491).
$C_2Cl_2O_2$	Oxalyl chloride, B 20° 0.92 (500).
C_2Cl_3N	Trichloroacetonitrile, Hx 2.0 (336).
C_2Cl_4	Tetrachloroethylene, B 25° 0 (512).
C_2Cl_4O	Trichloroacetyl chloride, B 20° 1.19 (498).
$C_2Cl_4O_2$	Trichloromethyl chloroformate, Hx 2.16 (611).
C_2Cl_6	Hexachloroethane, B or Hx 15-30° 0 (546); Hx 25° 0 (556).
$C_2F_6N_2$	"Nitrogen-carbon-fluorine compound," 301-368°K 0.47 (445).

C_2HBr Bromoacetylene, 289-354°K $\underline{0}$ (581).

C_2HBr_3O Tribromoacetaldehyde, B 20° $\underline{1.69}$ (583).

C_2HCl Chloroacetylene, 287-363°K $\underline{0.44} \pm .01$ (581).

C_2HCl_2N Dichloroacetonitrile, Hx $\underline{2.5}$ (336); B or CT 25° $\underline{2.0}$ (390).

C_2HCl_3 Trichloroethylene, CT $\underline{0.8}$ (208); B 25° $\underline{0.94}$ (221).

C_2HCl_3O Trichloroacetaldehyde, B 20° $\underline{1.58}$ (583).

$C_2HCl_3O_2$ Trichloroacetic acid, B 25° $\underline{0.825}$ (603).

C_2H_2 Acetylene, 196-461°K $\underline{0}$ (14); 298°K $\underline{0}$ (394).

C_2H_2BrCl 1,2-Chlorobromoethylene, cis-, S $\underline{1.54}$ (39); S $\underline{1.55}$ (657); trans-, S $\underline{0}$ (657).

$C_2H_2Br_2$ 1,2-Dibromoethylene, cis-, B $\underline{1.35}$ (12); S $\underline{1.22}$ (39); trans-, B $\underline{0}$ (12).

$C_2H_2Br_4$ 1,1,2,2-Tetrabromoethane, Hx 25° $\underline{1.31}$ (556).

C_2H_2ClI 1,2-Chloroiodoethylene, cis-, B $\underline{0.57}$ (38); trans-, B $\underline{1.27}$ (38).

C_2H_2ClN Chloroacetonitrile, B or CT 25° $\underline{2.86}$ (274); Hx $\underline{3.0}$ (336); B, CT or Hx 25° $\underline{2.97}$ (390); B 25° $\underline{3.00}$ (857).

$C_2H_2Cl_2$ 1,1-Dichloroethylene, B 25° $\underline{1.30}$ (858).

    1,2-Dichloroethylene, cis-, B $\underline{1.89}$ (12); S $\underline{1.85}$ (39); B 25° $\underline{1.74}$ (127); S $\underline{1.74}$ (657); Vap. $\underline{2.95}$ (735); trans-, B $\underline{0}$ (12); B 25° $\underline{0.70}$ (127); S $\underline{0}$ (657).

$C_2H_2Cl_2O$ Chloroacetyl chloride, 358-529°K $\underline{2.2}$ (271); 20° B $\underline{2.22}$, CD $\underline{2.06}$ (498).

$C_2H_2Cl_2O_2$ Dichloroacetic acid, B 25° $\underline{1.090}$ (603).

$C_2H_2Cl_4$ 1,1,1,2-Tetrachloroethane, B or CT $\underline{1.2}$ (208).

    1,1,2,2-Tetrachloroethane (acetylene tetrachloride), B $\underline{1.95}$, CT $\underline{1.85}$ (208); 401-436°K $\underline{1.36}$ (449); 0-50° B $\underline{1.70}$, Hx $\underline{1.42}$, E $\underline{1.95}$ (502); Hx 25° $\underline{1.44}$ (556).

$C_2H_2I_2$ 1,2-Diiodoethylene, cis-, B $\underline{0.71}$ (12); S $\underline{0.75}$ (39); trans-, B $\underline{0}$ (12).

C_2H_2O Ketene, 398-446°K $\underline{1.45}$ (831).

C_2H_3Br Vinyl bromide, 295-413°K $\underline{1.407} \pm .005$ (594).

C_2H_3BrO Acetyl bromide, B 20° $\underline{2.43}$ (498).

C_2H_3Cl Vinyl chloride, 287-413°K $\underline{1.442} \pm .01$ (594).

C_2H_3ClO Acetyl chloride, 320-483°K $\underline{2.68}$ (271); B 20° $\underline{2.45}$ (498); B 25° $\underline{2.40}$ (545).

$C_2H_3ClO_2$ Methyl chloroformate, B $\underline{2.22}$ (611).

    Chloroacetic acid, B 25° $\underline{1.524}$ (603); B 30° $\underline{2.29}$ (699).

$C_2H_3Cl_3$ 1,1,1-Trichloroethane, B 25° $\underline{1.57}$ (190); B or CT $\underline{1.5}$ (208); Vap. $\underline{2.03}$ (735); 336-399°K $\underline{1.77}$ (748).

    1,1,2-Trichloroethane, B $\underline{1.55}$, CT $\underline{1.15}$ (208).

C_2H_3I Vinyl iodide, 290-413°K $\underline{1.26} \pm .015$ (594).

C_2H_3N Acetonitrile, B 25° $\underline{3.4}$ (62); B 20° $\underline{3.11}$ (97); B 18° $\underline{3.51}$ (111); B 20° $\underline{3.16}$ (238); B or CT 25° $\underline{3.42}$ (274); Hx $\underline{3.4}$ (336); B, CT or Hx 25° $\underline{3.45}$ (390); B 20° $\underline{3.44} \pm .02$ (411, 441); 354-463°K $\underline{3.94} \pm .01$ (536); B 25° $\underline{3.51}$, T -60° $\underline{3.18}$, -30° $\underline{3.26}$, 0° $\underline{3.32}$, 30° $\underline{3.40}$, 60° $\underline{3.45}$ (653); 25° gasoline $\underline{3.33}$, kerosene or petr. ether $\underline{3.35}$ (853).

C_2H_3NS Methyl isothiocyanate, B 20° $\underline{3.18}$ (238).

    Methyl thiocyanate, B 20° $\underline{3.16}$ (238).

$C_2H_3N_3$ 1,2,3-Triazole, B 25° $\underline{1.77}$ (787).

    1,2,4-Triazole, D 20° $\underline{3.23}$, 50° $\underline{3.16}$, 96° $\underline{3.47}$ (680); D 25° $\underline{3.24}$ (727); D 25° $\underline{3.17}$ (787).

$C_2H_3N_3O$ 1,2,4-Triazole-5-one, D 25° $\underline{3.30}$ (787).

C_2H_4 Ethylene, 237-461°K $\underline{0}$ (14); 298°K $\underline{0}$ (394).

C_2H_4BrCl 1-Chloro-2-bromoethane (ethylene chlorobromide), Hp -50° $\underline{0.92}$, 30° $\underline{1.19}$ (182); Hp -50° $\underline{0.92}$, -30° $\underline{0.98}$, -10° $\underline{1.05}$, 10° $\underline{1.11}$, 30° $\underline{1.19}$, 50° $\underline{1.18}$, 70° $\underline{1.19}$ (183); 339°K $\underline{1.09}$, 368° $\underline{1.14}$,

405° $\underline{1.20}$, 436° $\underline{1.28}$ (269); 25° B $\underline{1.52}$, Am $\underline{1.17}$ (355); Hx -50° $\underline{0.85}$, -25° $\underline{0.95}$, 0° $\underline{1.04}$, 25° $\underline{1.14}$, 50° $\underline{1.21}$ (356).

$C_2H_4Br_2$ 1,1-Dibromoethane, B 25° $\underline{2.12}$ (81).
1,2-Dibromoethane, B 25° $\underline{1.4}$ (62); B 10° $\underline{1.46}$, 30° $\underline{1.52}$, 50° $\underline{1.52}$, 70° $\underline{1.55}$, Hp -30° $\underline{0.79}$, -10° $\underline{0.94}$, 10° $\underline{0.98}$, 30° $\underline{1.04}$, 50° $\underline{1.05}$, 70° $\underline{1.05}$ (184); Debye eq. 347°K $\underline{0.97}$, 449° $\underline{1.04}$, Meyer eq. 357° $\underline{0.99}$, 455° $\underline{1.15}$ (229); 339°K $\underline{0.94}$, 368° $\underline{0.99}$, 405° $\underline{1.03}$, 436° $\underline{1.10}$ (269); Hx -50° $\underline{0.57}$, -25° $\underline{0.67}$, 0° $\underline{0.79}$, 25° $\underline{0.91}$, 50° $\underline{0.92}$ (356); B 10° $\underline{1.13}$, 20° $\underline{1.15}$, 30° $\underline{1.18}$, 40° $\underline{1.20}$, T 10° $\underline{0.95}$, 20° $\underline{0.98}$, 30° $\underline{1.01}$, 40° $\underline{1.03}$, CT 10° $\underline{0.86}$, 20° $\underline{0.89}$, 30° $\underline{0.92}$, 40° $\underline{0.95}$, CH 10° $\underline{0.84}$, 20° $\underline{0.88}$, 30° $\underline{0.91}$, 40° $\underline{0.93}$ (379); Hx 25° $\underline{0.91}$ (556); 309°K $\underline{1.002}$, 339° $\underline{1.04}$, 369° $\underline{1.08}$ (712).

$C_2H_4ClNO_2$ 1-Chloro-1-nitroethane, 415-468°K $\underline{3.33}$ (758).

$C_2H_4Cl_2$ 1,1-Dichloroethane, 310-413°K $\underline{2.045}$ (76); B 25° $\underline{1.95}$ (81); B 25° $\underline{1.98}$ (166); B or CT $\underline{1.8}$ (208); Vap. $\underline{2.63}$ (735).
1,2-Dichloroethane (ethylene chloride), B 25° $\underline{1.75}$ (62); 305-392°K $\underline{1.567}$ (76); B 25° $\underline{1.87}$ (166); 333-453°K $\underline{1.2}$ to $\underline{1.4}$ (178); Hp -70° $\underline{1.07}$, 30° $\underline{1.41}$ (182); Hp -50° $\underline{1.16}$, -30° $\underline{1.24}$, -10° $\underline{1.31}$, 10° $\underline{1.36}$, 30° $\underline{1.41}$, 50° $\underline{1.42}$ (183); 305-544°K $\underline{1.12}$ to $\underline{1.54}$ (207); B $\underline{1.9}$, CT $\underline{1.3}$ (208); Debye eq. 298-588°K $\underline{1.27}$ to $\underline{1.57}$, Meyer eq. 298-588° $\underline{1.05}$ to $\underline{1.56}$ (229); 213°K $\underline{1.23}$, 293° $\underline{1.51}$ (246); E $\underline{1.5}$ (298); B 20° $\underline{1.71}$, 40° $\underline{1.69}$, 50° $\underline{1.68}$, Am -50° $\underline{1.22}$, -25° $\underline{1.29}$, 0° $\underline{1.33}$, 25° $\underline{1.38}$ (355); Hx -50° $\underline{1.13}$, -25° $\underline{1.21}$, 0° $\underline{1.30}$, 25° $\underline{1.36}$, 50° $\underline{1.42}$ (356); E 20° $\underline{1.5}$ (358); B 10 to 70° $\underline{1.77}$ to $\underline{1.74}$, CT -20 to 60° $\underline{1.33}$ to $\underline{1.45}$, E -20 to 20° $\underline{1.26}$ to $\underline{1.44}$, CD -70 to 30° $\underline{1.22}$ to $\underline{1.40}$, Hp -10 to 90° $\underline{1.31}$ to $\underline{1.57}$, Cf -50 to 50° $\underline{1.25}$ to $\underline{1.42}$ (391); CT -18 to 46° $\underline{1.53}$ to $\underline{1.59}$ (504); Hx 25° $\underline{1.36}$ (556); 308°K $\underline{1.43}$, 335° $\underline{1.47}$, 372° $\underline{1.51}$, D 25° $\underline{1.84}$ (712).

$C_2H_4I_2$ 1,1-Diiodoethane, B 25° $\underline{2.30}$ (81).
1,2-Diiodoethane, B 25° $\underline{1.3}$ (62); Hx 25° $\underline{0.44}$, 50° $\underline{0.55}$ (356).

$C_2H_4N_2O_2$ Glyoxime, D 20° $\underline{1.22}$ (366).

$C_2H_4N_4$ 1-Methyl-1,2,3,4-tetrazole, B 25° $\underline{5.38}$ (787).

C_2H_4O Acetaldehyde, 300-455°K $\underline{2.69}$ (271); B 20° $\underline{2.49}$ (561, 583); 420-469°K $\underline{2.72}$ (783).
Ethylene oxide, 290-449°K $\underline{1.88}$ (57); S $\underline{1.88}$ (236).

$C_2H_4O_2$ Acetic acid, B 22° $\underline{1.04}$ (105); Lqd. 10-70° $\underline{0}$ (137); B 25° $\underline{0.74}$ $\pm$ .02 (147); 411°K $\underline{1.4}$, 471°K $\underline{1.7}$ (150); 298-494°K $\underline{1.73}$ (206); E -60° $\underline{0.7}_2$, -40° $\underline{0.8}_5$, -20° $\underline{0.9}_3$, 0° $\underline{1.0}$, 15° $\underline{1.1}_4$, 25° $\underline{1.24}$ (299, 358); Hx 10° $\underline{0.37}_8$, 30° $\underline{0.38}_4$ (377); D 25° $\underline{1.74}$ (399); B 30° $\underline{1.77}$ (699); Vap. $\underline{1.73}$, D 23° $\underline{1.74}$ (819); D $\underline{1.75}$ (866).

C_2H_5Br Ethyl bromide, 303-390°K $\underline{1.779}$ (45); Lqd. -90 to 60° $\underline{1.86}$ (51); B $\underline{2.12}$ (71); 301-419°K $\underline{1.789}$ (124); B 25° $\underline{1.77}$ (372); 236-535°K $\underline{1.92}$ (388); 301-420°K $\underline{1.99}$ (429); B 6-28° $\underline{1.94}$, T 0-20° $\underline{1.75}$ to $\underline{1.78}$, Hp 0-20° $\underline{1.96}$ (504); B 10-40° $\underline{1.88}$, T -23 to 40° $\underline{1.81}$ to $\underline{1.84}$, CT 0-40° $\underline{1.89}$, CD -23 to 20° $\underline{1.70}$, H -23 to 40° $\underline{1.55}$ to $\underline{1.59}$, CH 10-40° $\underline{1.56}$ (527); 292-443°K $\underline{2.01}$ $\pm$ .01 (536); CT 20° $\underline{1.99}$ (711).

C_2H_5Cl Ethyl chloride, 293°K Kerr $\underline{1.76}$ (25); 297-474°K $\underline{1.98}$ (50); 298-418°K $\underline{2.00}$ $\pm$ .01 (87); 292-455°K $\underline{2.05}$ (114); 304-421°K $\underline{1.997}$ (124); 298-418°K $\underline{2.02}_0$ $\pm$ .025 (133, 254); B or CT $\underline{1.8}$ (208); 304-421°K $\underline{2.00}$ (429); T 0-21° $\underline{1.79}$ (504); 292-359°K $\underline{1.98}$ $\pm$ .03 (851).

C_2H_5ClO Chloromethyl methyl ether, CT 10-40° $\underline{1.88}$ (380).
Ethylene chlorohydrin, B 25-50° $\underline{1.89}$, Hp 25-50° $\underline{2.08}$ (259); 339-435°K $\underline{1.7}_4$ (270).

$C_2H_5ClO_2S$	Ethanesulfonyl chloride, B 25° $\underline{4.86}$ (848).
	Ethyl chlorosulfite, B 25-45° $\underline{2.63}$ to $\underline{2.66}$ (455).
C_2H_5F	Ethyl fluoride, 236-535°K $\underline{1.92}$ (388).
C_2H_5I	Ethyl iodide, 320-398°K $\underline{1.620}$ (45); B 25° $\underline{1.7}$ (62); S $\underline{1.66}$ (89); 306-404°K $\underline{1.623}$ (124); 348-463°K $\underline{1.90}$ (388); 306-404°K $\underline{1.93}$ (429); B 6-28° $\underline{1.82}$, T 0-24° $\underline{1.69}$, Hp 0-20° $\underline{1.74}$ (504); B 20° $\underline{1.78}$ (528); 293-337°K $\underline{1.87} \pm .01$ (536); CT 20° $\underline{1.89}$ (711, 750).
C_2H_5NO	Acetamide, D 20° $\underline{3.6}$ (284); D 30° $\underline{3.72}$ (363).
$C_2H_5NO_2$	Aminoacetic acid (glycine), water 1-30° $\underline{20.8}$ (605).
	Ethyl nitrite, B or CT 25° $\underline{2.3}$ (274); B 20° $\underline{2.20}$ (279); CT 0° $\underline{2.29}$ (334, 336, 390); 290°K $\underline{2.38}$ (344).
	Nitroethane, B 20° $\underline{3.19}$ (303); 365-461°K $\underline{3.58} \pm .01$ (536); 416-484°K 3.70 (758).
$C_2H_5NO_3$	Ethyl nitrate, B 20° $\underline{2.91}$ (279).
C_2H_6	Ethane, 200-470°K $\underline{0}$ (14); 298°K $\underline{0}$ (394).
$C_2H_6N_2$	Azomethane, Hp 25° $\underline{0}$ (626).
$C_2H_6N_2O$	Nitrosodimethylamine, B 20° $\underline{3.98}$ (280).
$C_2H_6N_2S$	Methylthiourea, D 18° $\underline{4.2} \pm .1$ (578).
C_2H_6O	Ethyl alcohol, Lqd. 18° $\underline{0.53}$ (4); CT 25° $\underline{1.63}$ (24); 361-453°K $\underline{1.1}_1$ (26); S $\underline{1.58}$ (39); B 10-70° $\underline{1.74}$ (55); 351-499°K $\underline{1.696}$ (83); B 22° $\underline{1.70}_7 \pm .02$ (203); 297-450°K $\underline{1.68}_6$ (239); E $\underline{1.8}$ (298); E -60° $\underline{1.5}$, 0° $\underline{1.69}$, 20° $\underline{1.8}$ (299, 358); 2,5-dimethylpyrazine 25° $\underline{1.86}$ (334, 336, 390); B $\underline{1.74}$, E $\underline{1.8}$ (357); 308-483°K $\underline{1.67}$ (424); 25° B $\underline{1.700} \pm .006$, CT $\underline{1.674} \pm .005$ (477); CD $\underline{1.51}$ (541); B 30° $\underline{1.66}$ (734); Lqd. 20° $\underline{2.92}$, 70° $\underline{2.84}$ (816).
	Methyl ether, 292-453°K $\underline{1.29}$ (57); 298-418°K $\underline{1.32} \pm .02$ (86); Gas $\underline{1.316} \pm .012$ (133); 218-378°K $\underline{1.28}_7 + .01$ (254); 290-428°K $\underline{1.28} \pm .01$ (537).
$C_2H_6O_2$	Ethylene glycol, B 25° $\underline{1.5}$ (62); D 25° $\underline{2.28}$, 50° $\underline{2.30}$ (186); 417-506°K $\underline{2.2}_5$ (270); E -60° $\underline{1.24}$, -40° $\underline{1.6}_9$, -20° $\underline{1.9}_0$, 20° $\underline{2.3}_8$ (299, 358); D 15° $\underline{2.18} \pm .02$, 30° $\underline{2.20} \pm .02$ (707); B 30° $\underline{2.30}$ (780); Lqd. 20° $\underline{3.49}$ (816).
$C_2H_6O_2S$	Methyl sulfone, 424-526°K $\underline{4.41} \pm .1$ (636).
$C_2H_6O_2S_2$	Dimethyl thiosulfite, B $\underline{1.89} + .02$ (619).
$C_2H_6O_3S$	sym-Dimethyl sulfite, B 20° $\underline{2.90}$ (303).
$C_2H_6O_4S$	Dimethyl sulfate, S $\underline{3.27}$ (39).
C_2H_6S	Ethyl mercaptan, 308-478°K $\underline{1.56}$ (485); B 15° $\underline{1.38} \pm .02$ (707).
	Methyl sulfide, B 20° $\underline{1.40}$ (238).
C_2H_7N	Dimethylamine, 298-418°K $\underline{0.96} \pm .01$ (139, 187, 254); 296-321°K $\underline{0.90}$ (165); 288-427°K $\underline{1.02}$ (537); 298°K $\underline{1.02}$, B 25° $\underline{1.17}$, Lqd. 0 to 25° $\underline{1.13}$ (852).
	Ethylamine, 295-344°K $\underline{0.99}$ (165); B 25° $\underline{1.39}$ (511); B 25° $\underline{1.37}$ (669).
C_2H_7NO	2-Aminoethanol (ethanolamine), D 25° $\underline{2.27}$ (443).
$C_2H_7NO_2S$	Ethanesulfonamide, 25° B $\underline{4.03}$, D $\underline{4.62}$ (849).
$C_2H_8N_2$	Ethylenediamine, 355-429°K $\underline{1.9}_4$ (270); B 25-75° $\underline{1.92}$ (511, 669).
	N,N'-Dimethylhydrazine, Hp 25° $\underline{1.35}$ (626).
C_2I_2	Diiodoacetylene, CT 0° ca. $\underline{0.33}$ (322).
C_2N_2	Cyanogen, 293-415°K $\underline{0.3}$ (158); 193-298°K $\underline{0}$ (513).
C_3Cl_6	Hexachloropropylene, CT $\underline{0.45}$ (208).
C_3Cl_8	Octachloropropane, CT $\underline{0}$ (208).
C_3HCl_7	1,1,1,2,2,3,3-Heptachloropropane, CT $\underline{1.0}$ (208).
$C_3H_2Cl_2O_2$	Malonyl chloride, B 20° $\underline{2.80}$ (500).
$C_3H_2N_2$	Propanedinitrile, B 25° $\underline{3.56}$, 75° $\underline{3.61}$ (569, 669).
$C_3H_2O_2$	Propiolic acid, D 25° $\underline{2.08}$ (460).

C_3H_3N Acrylonitrile, 387-509°K $\underline{3.88}$ (783); B 25° $\underline{3.51}$ (859).

C_3H_3NO Isoxazole, B 25° $\underline{2.81}$ (787).

$C_3H_3NOS_2$ 4-Oxo-2-thionthiazolidene (rhodanine), D 25° $\underline{2.20}$ (787).

$C_3H_3NO_2S$ 2,4-Dioxothiazolidene, D 25° $\underline{2.03}$ (787).

C_3H_3NS Thiazole, B 25° $\underline{1.64}$ (787).

C_3H_4 Allene, 298-368°K $\underline{0.20}$ (513).

Allylene (methylacetylene), 298-368°K $\underline{0.72}$ (513); 298-348°K $\underline{0.77}$ (596).

$C_3H_4Cl_2$ 1,1-Dichlorocyclopropane (cyclopropylidene chloride), B 25° $\underline{2.04}$ (841).

1,2-Dichlorocyclopropane, dl-, B 25° $\underline{1.18}$ (844).

1,1-Dichloropropene, B 25° $\underline{1.73}$ (221); B 25° $\underline{1.69}$ (858).

$C_3H_4N_2$ Imidazole, B 50° $\underline{6.21}$, 60° $\underline{6.17}$, 70° $\underline{6.16}$, D 20° $\underline{4.84}$, 50° $\underline{4.78}$, 90° $\underline{4.58}$ (680); 25° B $\underline{6.2}$, D $\underline{4.9}$ (727); B 25° $\underline{3.84}$ (787).

Pyrazole, B 20° $\underline{1.46}$, 50° $\underline{1.64}$, 70° $\underline{1.79}$, D 20° $\underline{2.19}$, CH 70° $\underline{1.32}$ (680); 25° B $\underline{1.47}$, $\underline{1.7}$, D $\underline{2.2}$ (727); B 25° $\underline{1.57}$ (787).

$C_3H_4N_2S$ 2-Aminothiazole, B 25° $\underline{1.75}$ (787).

C_3H_4O Acrylic aldehyde (acrolein), B 20° $\underline{2.88}$ (583); 377-478°K $\underline{3.04}$ (831).

C_3H_5Br Allyl bromide, B 20° $\underline{1.79}$ (175).

1-Bromopropene-1, cis-, B 25° $\underline{1.57}$ (858).

2-Bromopropene, B 25° $\underline{1.51}$ (858).

C_3H_5BrO Bromoacetone, Hx 20° $\underline{2.38}$ (609).

$C_3H_5Br_3$ 1,2,3-Tribromopropane, B 25° $\underline{1.57}$, 50° $\underline{1.59}$, Hp 25° $\underline{1.48}$, 50° $\underline{1.51}$ (259).

C_3H_5Cl Allyl chloride, 304-420°K $\underline{2.006}$ (124); S $\underline{1.99}$ (136); 304-420°K $\underline{2.02}$ (429); 308-480°K $\underline{1.88}$ (547); 377-480°K $\underline{1.98}$ (830).

Chlorocyclopropane, B 25° $\underline{1.76}$ (844, 841).

1-Chloropropene-1, cis-, B 25° $\underline{1.65}$ (858).

2-Chloropropene, B 25° $\underline{1.53}$ (858).

1-Chloropropene, cis-, 345-476°K $\underline{1.71}$ (818, 830); trans-, 345-476°K $\underline{1.97}$ (818, 830).

2-Chloropropene, 339-469°K $\underline{1.69}$ (818, 830).

C_3H_5ClO Chloroacetone, 336-454°K $\underline{2.1}_7$ to $\underline{2.24}$ (271); Hx 20° $\underline{2.35}$ (609).

α-Epichlorohydrin, CT $\underline{1.8}$ (208).

Propionyl chloride, B 20° $\underline{2.61}$ (498); B 25° $\underline{2.48}$ (545).

C_3H_5N Ethyl cyanide (propionitrile), B 25° $\underline{3.4}$ (62); B 20° $\underline{3.34}$ (97); B 18° $\underline{3.66}$ (111); B 20° $\underline{3.57}$ (411); T -79 to 20° $\underline{3.35}$ to $\underline{3.48}$, Hx -23 to 20° $\underline{3.62}$ to $\underline{3.66}$, CH 20° $\underline{3.65}$, CD -79 to 20° $\underline{3.15}$ to $\underline{3.30}$ (469); 351-469°K $\underline{4.03}$ ± .01 (536); B 25° $\underline{3.56}$, 75° $\underline{3.57}$ (569, 669); 395-477°K $\underline{4.00}$ (783).

Ethyl isocyanide, B 25° $\underline{3.47}$ (248).

C_3H_5NO Ethyl isocyanate, B 20° $\underline{2.81}$ (442, 467).

C_3H_5NS Ethyl isothiocyanate, B 20° $\underline{3.31}$ (238).

Ethyl thiocyanate, B 20° $\underline{3.64}$ (238).

C_3H_6 Propene (propylene), 246-476°K $\underline{0.35}$ (308); 298°K $\underline{0}$ (394).

$C_3H_6Br_2$ 1,3-Dibromopropane (trimethylene bromide), Hp -30° $\underline{2.07}$, -10° $\underline{2.11}$, 10° $\underline{2.15}$, 30° $\underline{2.18}$, 50° 2.19 (184); B 25° 1.97, 50° $\underline{1.98}$ (259); Hp 25° $\underline{2.02}$, 50° $\underline{2.03}$ (320).

$C_3H_6ClO_2N$ 1-Chloro-1-nitropropane, 416-493°K $\underline{3.52}$ (758).

$C_3H_6Cl_2$ 1,1-Dichloropropane, B 25° $\underline{2.06}$ (166).

1,3-Dichloropropane, B 25° $\underline{2.24}$ (166); 374-485°K $\underline{2.07}$ (449).

2,2-Dichloropropane, B 25° $\underline{2.18}$ (166); B or CT 25° $\underline{2.0}$ (274); Vap. $\underline{2.63}$ (735).

$C_3H_6N_2O_2$ Methylglyoxime, D 20° $\underline{0.882}$ (366).

C_3H_6O Acetone, CT 25° $\underline{2.70}$ (24); 291-456°K $\underline{2.84}$ (57); Hx 15° $\underline{2.71}$ (66); B 18° $\underline{2.72}$ (77); B 18° $\underline{2.78}$, T 16° $\underline{2.81}$, Cf 18° $\underline{2.66}$, An

$19°$ $\underline{2.37}$ (116, 121); B $\underline{2.78}$, T $\underline{2.80}$, Cf $\underline{2.64}$, An $\underline{2.75}$ (122); B $22°$ $\underline{2.74}_0 \pm .02$ (203); $301-455°K$ $\underline{2.85}$ (271); B $\underline{2.74} \pm .015$ (293); B $\underline{2.76}$, Hx $\underline{2.81}$, CD $\underline{2.70}$, E $\underline{2.2}$ (357); E $20°$ $\underline{2.2}$ (358); Vap. $\underline{2.85}$ (637); gasoline $25°$ $\underline{2.8}$ (854).

Allyl alcohol, Vap. $\underline{1.63}$ (425).

Propionaldehyde, B $20°$ $\underline{2.54}$ (561, 583); $354-509°K$ $\underline{2.73}$ (783).

Propylene oxide, B $25°$ $\underline{1.88}$ (236, 333); B $25°$ $\underline{1.98}$ (859).

Trimethylene oxide, B $25°$ $\underline{2.01}$ (236, 333).

$C_3H_6O_2$ Ethyl formate, B $25°$ $\underline{1.94}$, $50°$ $\underline{1.93}$ (185); $292-435°K$ $\underline{1.92}$ (272); B 25 and $50°$ $\underline{1.94}$ (315).

Methyl acetate, CT $25°$ $\underline{1.67}$ (24); B $25°$ $\underline{1.7}_5$ (127); B $22°$ $\underline{1.74}_2 \pm .015$ (203); $327-516°K$ $\underline{1.67}$ (272); E $\underline{2.2}$ (298); Lqd. $30-40°$ $\underline{1.74}$ (815).

Propionic acid, B $22°$ $\underline{0.88}$ (105); B $25°$ $\underline{0.63} \pm .01$ (147); $356-486°K$ $\underline{1.74}$ (206); Hx $10°$ $\underline{0.63}$, $30°$ $\underline{0.65}$ (377); D $25°$ $\underline{1.75}$ (399); B $30°$ $\underline{1.68}$ (699); Vap. $\underline{1.74}$, D $23°$ $\underline{1.75}$ (819); D $\underline{1.50}$ (866).

$C_3H_6O_3$ Dimethyl carbonate, $328°K$ $\underline{0.86}$, $350°$ $\underline{0.89}$, $412°$ $\underline{0.94}$, $479°$ $\underline{1.00}_5$ (548); B $25°$ $\underline{1.06}$ (668); T $10°$ $\underline{0.90}$, $25°$ $\underline{0.95}$, $50°$ $\underline{1.02}$, Hx $-25°$ $\underline{0.55}$, $0°$ $\underline{0.62}$, $25°$ $\underline{0.73}$, $50°$ $\underline{0.79}$ (675).

1,3,5-Trioxane, B $30°$ $\underline{2.18}$ (801).

C_3H_7Br Isopropyl bromide, B $\underline{2.20}$ (71); B $20°$ $\underline{2.09}$ (175); B $20°$ $\underline{2.04}$ (528); $287-380°K$ $\underline{2.19} \pm .01$ (536).

n-Propyl bromide, B $\underline{2.00}$ (71); $302-416°K$ $\underline{1.78}_9$ (124); S $\underline{1.78}$ (136); B $20°$ $\underline{1.94}$ (175); $302-416°K$ $\underline{2.01}$ (429); B $20°$ $\underline{1.93}$ (528); $348-441°K$ $\underline{2.15} \pm .01$ (536); B $\underline{1.97}$, T $\underline{1.93}$, Hx $\underline{2.07}$, Hp $\underline{2.06}$, CT $\underline{1.99}$ (840).

C_3H_7BrO Trimethylene bromohydrin, B $25°$ $\underline{2.17}$, $50°$ $\underline{2.21}$ (259).

C_3H_7Cl Isopropyl chloride, B $20°$ $\underline{2.04}$ (175); $288-383°K$ $\underline{2.15} \pm .01$ (536). n-Propyl chloride, $338-458°K$ $\underline{1.90} \pm .01$ (87); $304-402°K$ $\underline{2.028}$ (124); $338-458°K$ $\underline{2.04}_0 \pm .007$ (133, 178, 254); S $\underline{2.07}$ (136); B $20°$ $\underline{1.94}$ (175); B $\underline{1.92}$, Hx $\underline{1.99}$, CD $\underline{1.81}$, E $\underline{1.7}$ (357); $304-402°K$ $\underline{2.04}$ (429); B $0-40°$ $\underline{1.97}$, Hp $0-40°$ $\underline{2.02}$, CT $0-40°$ $\underline{1.92}$ to $\underline{1.96}$, CD $0-40°$ 1.81 to 1.84 (504).

C_3H_7ClO Trimethylene chlorohydrin, B $25°$ $\underline{2.19}$, $50°$ $\underline{2.24}$, Hp $25°$ $\underline{2.30}$, $50°$ $\underline{2.35}$ (259).

$C_3H_7ClO_2S$ Isopropyl chlorosulfite, B $25°$ $\underline{2.66}$, $45°$ $\underline{2.68}$ (456). n-Propyl chlorosulfite, B $25°$ $\underline{2.71}$, $45°$ $\underline{2.73}$ (455).

C_3H_7I Isopropyl iodide, B $20°$ $\underline{1.99}$ (175); B $20°$ $\underline{1.84}$ (528); CT $20°$ $\underline{2.08}$ (750). n-Propyl iodide, $306-404°K$ $\underline{1.62}_7$ (124); B $20°$ $\underline{1.85}$ (175); $305-404°K$ $\underline{1.97}$ (429); B $20°$ $\underline{1.84}$ (528); $337-374°K$ $\underline{2.01} \pm .01$ (536); CT $20°$ $\underline{1.92}$ (750); B $\underline{1.84}$, T $\underline{1.83}$, Hx $\underline{1.94}$, Hp $\underline{1.92}$, CT $\underline{1.86}$ (840).

$C_3H_7NO_2$ 1-Nitropropane, $343-466°K$ $\underline{3.57} \pm .01$ (536); $382-457°K$ $\underline{3.72}$ (739, 748). 2-Nitropropane, $392-455°K$ $\underline{3.73}$ (739, 748). n-Propyl nitrite, B $20°$ $\underline{2.28}$ (279).

$C_3H_7NO_3$ n-Propyl nitrate, B $20°$ $\underline{2.98}$ (279).

C_3H_8 Propane, $227-486°K$ 0 (308); $298°K$ 0 (394).

$C_3H_8N_2O$ Dimethylurea, sym-, B $20°$ $\underline{5.1}$ (284); D $20°$ $\underline{4.8}$ (302).

C_3H_8O Isopropyl alcohol, B $10-70°$ $\underline{1.75}$ (55); S $\underline{1.78}$ (74); B $25°$ $\underline{1.78}$ (80); B $22°$ $\underline{1.69}_9 \pm .03$ (161); B or Hx $7-30°$ $\underline{1.63} \pm .02$ (419); $307-482°K$ 1.58 (424); B $30°$ 1.64 (430); $20°$ B 1.70_6, CD 1.48^* (475); B $25°$ $\underline{1.70}$ (476); $300-463°K$ $\underline{1.68}_2 \pm .007$ (566); B $30°$ $\underline{1.66}$ (734).

n-Propyl alcohol, B $24-70°$ $\underline{1.53}$, T $24°$ $\underline{1.53}$ (13); $376-505°K$ $\underline{1.65}_7$ (83); B $22°$ $\underline{1.65}_7 \pm .02$ (203); Vap. $\underline{1.64}$ (425); B $\underline{1.56}$

18

	(564); B <u>1.71</u> (655); Lqd. 20° <u>3.09</u> (816).
$C_3H_8O_2$	Dimethoxymethane, 307°K <u>0.74</u>, 329° <u>0.81</u>, 352° <u>0.84</u>, 383° <u>0.92</u>, 407° <u>0.97</u>, 472° <u>1.13</u> (486, 556).
	2-Methoxyethanol, B 25° <u>2.04</u> (633); B 30° <u>2.20</u> (780).
	1,2-Propanediol, D 25° <u>2.25</u>, 50° <u>2.28</u> (186); D <u>2.2</u> (266); Lqd. 20° <u>3.63</u> (816).
	1,3-Propanediol, D 25° <u>2.50</u>, 50° <u>2.51</u> (186); D <u>2.35</u> (266); Lqd. 20° <u>4.21</u> (816).
$C_3H_8O_3$	Glycerol, Lqd. 18° <u>0.28</u> (4); D 15° <u>2.67</u> ± .02, 30° <u>2.66</u> ± .02 (707).
C_3H_8S	n-Propyl mercaptan, B 20° <u>1.33</u> (238).
C_3H_9N	Trimethylamine, 338–458°K <u>0.60</u> ± .02 (139, 187, 254); 295–364°K <u>0.82</u> (165); 289–418°K <u>0.65</u> (537); 298°K <u>0.64</u>, B 25° <u>0.86</u>, Lqd. -10 to 25° <u>0.72</u> (852).
C_3H_9NO	Trimethylamine oxide, D 25° <u>5.04</u>, B 45° <u>5.02</u> (689); B 25° <u>4.87</u> ± .15 (817).
$C_3H_{10}N_2$	1,3-Propanediamine, B 25° <u>1.94</u>, 45° <u>1.95</u> (511, 669).
C_3O_2	Carbon suboxide, B 25° <u>0.7</u> (427).
$C_4H_2Cl_2N_2$	2,5-Dichloropyrimidine, D 35° <u>2.27</u> (867).
$C_4H_2Cl_6O_2$	2,2,3,5,5,6-Hexachloro-1,4-dioxane, S <u>0</u> (354, 406).
$C_4H_2N_2$	Fumaronitrile, B 25° <u>0</u> (712).
$C_4H_3BrO_3$	α-Bromotetronic acid, D 25° <u>6.00</u> (684).
$C_4H_3ClN_2S$	2-Mercapto-5-chloropyrimidine, D 35° <u>0</u> (867).
$C_4H_3ClO_3$	α-Chlorotetronic acid, D 25° <u>5.69</u> (684).
$C_4H_3IO_3$	α-Iodotetronic acid, D 25° <u>5.59</u> (684).
$C_4H_3NO_2S$	2-Nitrothiophene, B 25° <u>4.12</u> (768).
$C_4H_3NO_5$	α-Nitrotetronic acid, D 25° <u>6.10</u> (761).
$C_4H_4Cl_2O_4$	α,β-Dichlorosuccinic acid, dl-, B <u>2.93</u> (118); meso-, B <u>2.47</u> (118).
$C_4H_4Cl_4O_2$	2,3,5,6-Tetrachloro-1,4-dioxane, MP64° S <u>1.85</u>, MP100° S <u>1.05</u> (354); MP60°S <u>0</u>, MP100° S <u>1.85</u>, MP144° S <u>1.05</u> (406).
$C_4H_4N_2$	Pyrazine, Vap. <u>0</u> (273); B <<u>1</u> (335); D 35° <u>0.66</u> (867).
	Pyridazine, D 35° <u>3.94</u> (867).
	Pyrimidine, D 35° <u>2.42</u> (867).
	Succinonitrile, B 25° <u>3.8</u> (62); B 25° and 75° <u>3.93</u> (569, 669); T -90° <u>2.94</u>, -60° <u>3.16</u>, -30° <u>3.36</u>, 0° <u>3.54</u>, 30° <u>3.68</u>, 60° <u>3.80</u>, 90° <u>3.90</u> (653); 443°K <u>3.57</u>, 25° B <u>3.90</u>, D <u>3.80</u> (712).
$C_4H_4N_2O$	4-Oxypyrimidine, D 35° <u>2.70</u> (867).
	Pyridazone, D 25° <u>2.69</u> (727).
C_4H_4O	Furan, B 25° <u>0.71</u> (260); B 20° <u>0.67</u> ± .02 (660).
$C_4H_4O_2$	Diketene, 25° B <u>3.16</u>, CT <u>3.30</u> (401); B 25° <u>3.31</u> (770); 433–516°K <u>3.53</u> (783).
	Tetrolic acid, D 25° <u>2.12</u> (460).
$C_4H_4O_3$	Succinic anhydride, D 10° <u>4.16</u>, 20°–40° <u>4.20</u> to <u>4.22</u> (563).
	Tetronic acid, D 25° <u>4.72</u> (684).
C_4H_4S	Thiophene, S <<u>0.63</u> (117); B or Hx <u>0.53</u> to <u>0.54</u> (237); 329–474°K <u>0.58</u> (485); B 20° <u>0.54</u> ± .02 (660).
C_4H_4Se	Selenophene, B 20° <u>0.41</u> ± .03 (660); 25° B <u>0.78</u>$_1$, Hx <u>0.77</u>$_2$ (664).
C_4H_5Cl	2-Chloro-1,3,-butadiene (chloroprene), S <u>1.42</u> (685).
	4-Chloro-1,2,-butadiene, 394–491°K <u>2.02</u> (830).
C_4H_5ClO	Isocrotyl chloride, 358–523°K <u>1.99</u> (783).
$C_4H_5Cl_3O_2$	Ethyl trichloroacetate, B 25° <u>2.55</u> (603).
C_4H_5N	Crotononitrile (trans-), 409–516°K <u>4.50</u> (783).
	Cyclopropyl cyanide, B 25° <u>3.75</u> (841).
	Methacrylonitrile, 395–473°K <u>3.69</u> (831).
	Pyrrole, B 20° <u>1.83</u> (282); S <u>1.78</u> (393); B 20° <u>1.80</u> ± .01 (660);

19

B 20° 2.1, 22° 1.8, 23° 1.7 (680).

C$_4$H$_5$NO	α-Methylisoxazole, B 3.13 (704).
	γ-Methylisoxazole, B 2.86 (704).
C$_4$H$_5$NO$_2$	Succinimide, B 20° 1.54 (468).
C$_4$H$_5$NS	Allyl isothiocyanate, B 20° 3.30 (238).
C$_4$H$_6$	1,3-Butadiene, 299-462°K 0 (781).
	Ethylacetylene, 298-348°K 0.80 (596).
C$_4$H$_6$Br$_2$	1,4-Dibromo-2-butene, trans-, B 25° 1.63 (845).
C$_4$H$_6$Cl$_2$	1,1-Dichloro-2-methylpropene, B 25° 1.73 (221).
	2,3-Dichloro-2-butene, cis-, B 21-30° 2.41 (693); trans-, B 20-26° 0 (693).
C$_4$H$_6$Cl$_2$O$_2$	2,3-Dichloro-1,4-dioxane, S 1.6 (406).
	Ethyl dichloroacetate, B 25° 2.61 (603).
C$_4$H$_6$Cl$_2$O$_3$	2-Hydroxy-2-dichloromethyl-1,3-dioxolane (glycol monodichloro-acetate (ring)), B 25° 3.35 (333).
C$_4$H$_6$N$_2$	4-Methylimidazole, B 20° 6.4, 50° 6.3, 70° 6.2, D 20° 5.1, GT 18° 5.8 (680); 25° B 6.3, D 5.1, CT 5.8 (727).
	N-Methylimidazole, 20° B 3.63, D 3.8 (680).
	1-Methylpyrazole, B 25° 2.28 (787).
	3-Methylpyrazole, B 25° 1.43 (787).
C$_4$H$_6$N$_2$O	Dimethylfurazan, S 4.008 (433).
	Dimethyloxydiazole, S 3.29 (433).
C$_4$H$_6$N$_2$O	3-Methyl-5-pyrazolone, D 25° 2.54 (787).
C$_4$H$_6$N$_2$O$_2$	Dimethylfurazan peroxide, B 25° 4.77 (666).
	Dimethylfuroxan, B 25° 4.40 (787).
	Ethyl diazoacetate (diazoacetic ester), B 25° 2.03 ± .02 (147); B 22° 2.025 ± .015 (268).
C$_4$H$_6$O	Crotonaldehyde, 412-519°K 3.67 (783).
	2,5-Dihydrofuran, B 20° 1.53 ± .01 (660).
	3,4-Epoxybutene-1, B 25° 1.84 (859).
	Ethoxyacetylene, B 25° 1.98 (796).
	α-Methylacrolein (methacrolein), 366-466°K 2.68 (831); B 25° 2.72 (858).
	Methyl vinyl ketone, B 25° 2.98 (859).
	Vinyl ether, B 20° 1.06 (260).
C$_4$H$_6$O$_2$	γ-Butyrolactone, B 25° 4.12 (496).
	Diacetyl, 329°K 1.25, 359° 1.29, 391° 1.33, 426° 1.37, 461° 1.42, 504° 1.48 (269); 328°K 1.22, 362° 1.27 (712).
	Methacryclic acid, B, Hx or D 0-40° 1.175-1.79 (798).
	Vinyl acetate, S 1.75 (685).
C$_4$H$_6$O$_3$	Acetic anhydride, 320-540°K 2.7 to 2.9 (332); CD 25° 2.46 (377).
C$_4$H$_6$S	Divinyl sulfide, 400-461°K 1.20 (830).
C$_4$H$_7$BrO	Bromomethyl ethyl ketone, Hx 2.33 (609).
C$_4$H$_7$BrO$_2$	2-Bromomethyl-1,3-dioxolane, B 25° 2.28 (560).
C$_4$H$_7$Cl	2-Methyl-3-chloropropene (methallyl chloride), 377-477°K 1.85 (830).
C$_4$H$_7$ClO	n-Butyryl chloride, B 20° 2.61 (498); B 25° 2.49 (545')
C$_4$H$_7$ClO$_2$	Ethyl chloroacetate, B 25° 2.64 (603).
C$_4$H$_7$ClO$_3$	Glycol mono-chloroacetate, B 25° 3.94 (333).
C$_4$H$_7$N	n-Butyronitrile, B 20° 3.46 (97); B 20° 3.57 (411); 339-443°K 4.05 ± .01 (536).
	Isopropyl cyanide (isobutyronitrile), B 25° 3.61 (857).
	3-Pyrroline (2,5-dihydropyrrole), B 20° 1.42 ± .01 (660).
C$_4$H$_7$NO	Acetone cyanhydrin, B 25° 3.17 (857).
	2-Pyrrolidone (2-ketopyrrolidine), B 20° 2.3 (284).
C$_4$H$_8$	1-Butene (α-butylene), 274-466°K 0.37 (14); 298-368°K 0.30 (513).

2-Butene (β-butylene), trans-, 298-368°K $\underline{0}$ (513).

Isobutylene, 298-368°K $\underline{0}$ (513).

$C_4H_8Br_2$ 1,4-Dibromobutane, B 25° $\underline{2.00}$, 50° $\underline{2.03}$, Hp 25° $\underline{1.96}$, 50° $\underline{2.01}$ (259).

2,3-Dibromobutane, **dl**- and **meso**-, B 22° $\underline{2.20}$ (509); Lqd. 25° dl-, $\underline{2.06}$, **meso**-, $\underline{1.57}$ (710).

$C_4H_8Cl_2O$ α,β-Dichloroethyl **ether**, Hx 20° $\underline{1.76}$ (695).

β,β'-Dichloroethyl **ether**, B 25° $\underline{2.58}$, 50° $\underline{2.57}$ (259); Hx 20° $\underline{2.40}$ (695).

$C_4H_8Cl_2S$ β,β'-Dichloroethyl **sulfide** ("mustard gas"), Hx 20° $\underline{1.76}$ (557, 612).

$C_4H_8I_2O$ β,β'-Diiodoethyl **ether**, B 25° $\underline{2.23}$, 50° $\underline{2.23}$ (259).

$C_4H_8N_2$ Acetaldazine, Hp 25° $\underline{1.10}$ (626).

$C_4H_8N_2O_2$ Dimethyl glyoxime, D 20° $\underline{1.38}$ (366).

C_4H_8O n-Butyraldehyde, B 18° $\underline{2.46}$ (78); B 20° $\underline{2.57}$ (583); 354-412°K $\underline{2.72}$ (783).

Isobutyraldehyde, B 20° $\underline{2.58}$ (583).

Methyl ethyl ketone, B 15° $\underline{2.79}$ (66); B 22° $\underline{1.81}_5$ ± .03 (203); Lqd. 30-40° $\underline{3.2}$ (815).

Tetrahydrofuran, S $\underline{1.87}$ (236); B 25° $\underline{1.71}$, 50° $\underline{1.71}$ (260); B 20° $\underline{1.68}$ ± .01 (660).

C_4H_8OS 1,4-Thioxane, S $\underline{0.47}$ (354, 406).

C_4H_8OSe 1,4-Selenoxane, S $\underline{0.30}$ (406).

$C_4H_8O_2$ n-Butyric acid, B 22° $\underline{0.93}$ (105); Lqd. 10-70° $\underline{0}$ (137); B 16° $\underline{0.68}$ ± .02 (147); B 30° $\underline{1.9}$ (699); D $\underline{1.58}$ (866).

1,4-Dioxane, B 25° $\underline{0.45}$ (88); B 25° $\underline{0.4}$ (143); B 20° $\underline{0.40}$ (302); S $\underline{0.3}$ (354, 406); 337-487°K $\underline{0}$ (385); 329-479°K $\underline{0.43}$ (488); Lqd. 14-87° $\underline{0}$ (674).

Ethyl acetate, CT 25° $\underline{1.74}$ (24); B 25° $\underline{1.8}_1$ (127); Hp -70° $\underline{1.87}$, -50° $\underline{1.90}$, -30° $\underline{1.93}$, -10° $\underline{1.95}$, 30° $\underline{2.00}$ (183); B 25° $\underline{1.86}$, 50° $\underline{1.82}$ (185); B 22° $\underline{1.81}_5$ ± .03 (203); 302-467°K $\underline{1.76}$ (272); B 25 and 50° $\underline{1.86}$ (315); 25° B $\underline{1.81}$, Hx $\underline{1.82}$, CT $\underline{1.87}$ (334, 336, 390); S $\underline{1.86}$ (685).

Isobutyric acid, B, Hx or D 0-40° $\underline{1.175}$ to $\underline{1.79}$ (798).

2-Methyl-1,3-dioxolane, B 25° $\underline{1.21}$ (560).

Methyl propionate, B 22° $\underline{1.69}_3$ ± .03, Hp $\underline{1.65}_7$ ± .03, CT $\underline{1.73}_5$ ± .03 (203).

Propyl formate, B 22° $\underline{1.89}_3$ ± .03 (203).

$C_4H_8O_3$ Ethylene glycol monoacetate, B 30° $\underline{2.33}$ (780).

C_4H_8S Tetrahydrothiophene, B 20° $\underline{1.87}$ ± .01 (660).

C_4H_8Se Tetrahydroselenophene, B 20° $\underline{1.79}$ ± .01 (660).

C_4H_9Br n-Butyl bromide, S $\underline{1.87}$ (89); B 10-50° $\underline{1.97}$ (130); Hp -90 to 70° $\underline{1.81}$ (138); B 20° $\underline{1.88}$ (528); 352-474°K $\underline{2.15}$ ± .01 (536); CT 20° $\underline{1.95}$ (751); B $\underline{1.98}$, T $\underline{1.96}$, Hx $\underline{2.09}$, Hp $\underline{2.03}$, CT $\underline{1.98}$ (840).

sec-Butyl bromide, B 20° $\underline{2.12}$ (130); 343°K $\underline{2.20}$ (536).

tert-Butyl bromide, B 10-50° $\underline{2.21}$ (130); CT 20° $\underline{2.17}$ (751).

Isobutyl bromide, B 20° $\underline{1.97}$ (130).

C_4H_9Cl n-Butyl chloride, S $\underline{1.89}$ (89); B 10-50° $\underline{1.97}$ (130); Hp -90 to 70° $\underline{1.88}$ (138); 315-480°K $\underline{2.04}$ (450); 288-375°K $\underline{2.11}$ ± .01 (536).

sec-Butyl chloride, B 10-50° $\underline{2.09}$ (130); 336-392°K $\underline{2.12}$ (739, 748).

tert-Butyl chloride, B 20° $\underline{2.15}$ (130); Hp -70 to 70° $\underline{2.14}$ (180); CT $\underline{2.0}$ (208); B or CT 25° $\underline{2.1}$ (274); CT 20° $\underline{2.04}$ (711); 324-354°K $\underline{2.13}$ (748); CT 20° $\underline{1.90}$ (751); B $\underline{1.94}$, T $\underline{1.92}$, Hx $\underline{2.03}$, Hp $\underline{2.02}$, CT $\underline{1.95}$ (840).

Isobutyl chloride, B 10-50° 1.96 (130); 345-407°K 2.04 (739, 748).

C_4H_9ClO α-Chloroethyl ethyl ether, Hx 20° 1.81 (695).

β-Chloroethyl ethyl ether, Hx 20° 2.18 (695).

$C_4H_9ClO_2S$ n-Butyl chlorosulfite, B 25° 2.70, 45° 2.74 (455).

Isobutyl chlorosulfite, B 25° 2.83, 45° 2.82 (456).

C_4H_9I n-Butyl iodide, B 10-50° 1.88 (130); Hp -90 to 70° 1.59 (138); B 20° 1.88 (497); 349-415°K 2.08 (536); CT 20° 1.93 (750); B 1.93, T 1.90, Hx 2.00, Hp 1.98 (840).

sec-Butyl iodide, B 20° 2.04 (130); CT 20° 2.10 (750).

tert-Butyl iodide, B 20° 2.13 (130); CT 20° 2.20 (750).

Isobutyl iodide, B 20° 1.87 (130); CT 20° 1.92 (750).

C_4H_9N Pyrrolidine, B 20° 1.57 ± .01 (660).

C_4H_9NO N-Dimethylacetamide, D 30° 3.79 (363).

N-Ethylacetamide, D 30° 3.87 (363).

Ethyl acetimino ether, D 30° 1.33 (363).

Morpholine, B 20° 1.48 (615); B 25° 1.58 (652); B 30° 1.51 (692).

C_4H_9NOS Thionylisobutylamine, B 25° 1.62 (728).

$C_4H_9NO_2$ Glycine ethyl ester, B 5-75° 2.11 (330).

2-Methyl-2-nitropropane, 392-449°K 3.71 (748).

1-Nitro-n-butane, B 20° 3.29 (303); 373-470°K 3.35 ± .01 (536); B 25° 3.40 ± .1 (839).

tert-Nitrobutane, Vap. 3.71 (739).

$C_4H_9NO_3$ n-Butyl nitrate, B 20° 2.96 (279).

C_4H_{10} n-Butane, 298-368°K 0 (513).

Isobutane, 298-368°K 0 (513).

$C_4H_{10}N_2O$ n-Propyl urea, D 20° 4.1 (284).

$C_4H_{10}N_2O_2$ Diethyl hyponitrite, B 20° 1.5 (303).

$C_4H_{10}O$ n-Butyl alcohol, B 20° 1.65 (13); S 1.62 (74); B 25° 1.62 (82); 385-490°K 1.659 (83); B 20-70° 1.74 (90); D 25° 1.81, 50° 1.75 (186); B 22° 1.66₀ ± .02 (203); 308-479°K 1.63 (424); Vap. 1.59 (425); Lqd. 20-30° 2.93 (816).

sec-Butyl alcohol, B 30° 1.65 (430).

tert-Butyl alcohol, B 22° 1.66 ± .022 (161); Hp -50 to 70° 1.65 (180); B or Hx 7-30° 1.55 ± .02 (419); B 30° 1.66 (734); Lqd. 20° 1.74 (816).

Ethyl ether, Lqd. 18° 1.27 (4); -79 to 16° Lqd. or 80, 60 or 40% in B 1.20 (7); Lqd. 1.423 (8); B 20° 1.22 (13); CT 25° 1.24 (24); 316-433°K 0.9₉ (26); B 25° 1.22 (31); B -17 to 20° 1.04 (40); B 18° 1.22 (47); 289-455°K 1.14 (57); 313-433°K 1.10 ± .02 (86, 254); Vap. 291°K 1.14 (114); B 18° 1.27, Cf 19° 1.79 (116, 121); B 1.27, Cf 1.77 (122); B 1.14 (125); 313-433°K 1.14₆ ± .01₂ (134); B 20° 0.74 (421); 288-476°K 1.17 ± .01 (537); Vap. 1.13 (679).

Isobutyl alcohol, B 20° 1.72 (13); S 1.79 (74); B 25° 1.79 (80); E 1.8 (298, 358); B 1.72, E 1.8 (357); 328-481°K 1.63 (424); B 30° 1.65 (430); 20° B 1.70₂, CD 1.41 (475); B 25° 1.71 (476); B 1.365 (564); Lqd. 30° 2.93 (816).

$C_4H_{10}O_2$ 1,4-Butanediol, D 15° 2.40 ± .02, 30° 2.39 ± .02 (707).

α-Hydroxyethyl ethyl ether, Hx 20° 1.62 (695).

β-Hydroxyethyl ethyl ether (glycol monoethyl ether), B 25° 2.08 (633); Hx 20° 2.10 (695); B 30° 2.22 (780).

$C_4H_{10}O_2S$ Diethyl sulfoxalate, B 1.90 ± .02 (619).

Diethyl sulfone, B 25° 4.41 (322).

$C_4H_{10}O_2S_2$ Diethyl thiosulfite, B 2.01 ± .02 (619).

$C_4H_{10}O_3$ β,β'-Dihydroxyethyl ether, B 20° 2.31 (695).

$C_4H_{10}S$	n-Butyl mercaptan, B 20° 1.32 (238); B 25° 1.48, 50° 1.49 (326).
	Ethyl sulfide, B 25° 1.61 (322); B 25° 1.58, 50° 1.58 (326); 309-474°K 1.51 (485).
$C_4H_{11}N$	n-Butylamine, B 25° 1.40 (857).
	sec-Butylamine, B 25° 1.28 (857).
	tert-Butylamine, B 25° 1.29 (857).
	Diethylamine, B 25° 1.20 (163); 297-367°K 0.90 (165); 25° B 1.13₁, Hx 1.10₃ (539).
$C_4H_{11}NO_2$	Diethanolamine, D 25° 2.81 (443).
$C_4H_{12}N_2$	Tetramethylenediamine, B 25° 1.93, 45° 1.93 (511, 669).
$C_5H_2Br_3N$	2,4,6-Tribromopyridine, B 17-67° 2.05 (416).
$C_5H_3Br_2N$	2,6-Dibromopyridine, B 17-67° 3.43 (416).
	3,5-Dibromopyridine, B 17-67° 0.98 (416).
C_5H_4BrN	2-Bromopyridine, B 20-67° 2.98 (416).
	3-Bromopyridine, B 17-67° 1.93 (416).
C_5H_4ClN	4-Chloropyridine, B 25° 0.84 (814).
$C_5H_4O_2$	Furfural, B 3.57 (119); S 3.61 (620).
$C_5H_4O_3$	Citraconic anhydride, B 10° 4.23, 20° 4.27, 30° 4.28, 40° 4.32 (563).
$C_5H_5BrO_3$	Methyl α-bromotetronate, D 25° 6.19 (684).
C_5H_5ClIN	Pyridine iodochloride, B 25° 8.20 ± .01 (785).
$C_5H_5ClN_2O$	2-Methoxy-5-chloropyrimidine, D 35° 0 (867).
$C_5H_5IO_3$	Methyl α-iodotetronate, D 25° 5.59 (684).
C_5H_5N	4-Cyano-1,3-butadiene, 427-464°K 3.90 (831).
	Pyridine, B 24° 2.11 (13); B 15° 2.21 (211); B 10-40° 2.26 ±.01 (382); B 17-67° 2.23 (416); 20° B 2.26, Hx 2.21, CH 2.20, T 2.25, CT 2.33, CD 2.10 (606); 25° B 4.24, D 4.32 (689); 25° B 2.20, D 2.22 (814).
C_5H_5NO	4-Pyridol, B 50° 6.0 (814).
C_5H_6	1,3-Cyclopentadiene, B 0.45 (806); 344-452°K 0.53 (829).
$C_5H_6N_2$	2-Aminopyridine, B 17-67° 2.17 (416).
	3-Aminopyridine, B 17-67° 3.19 (416).
	4-Aminopyridine, B 17° 3.79 (416); D 25° 4.36 (814).
	Glutaronitrile, B 25° 3.91, 75° 3.99 (569, 669).
C_5H_7N	2-Methylcyclopropanecarbonitrile, B 25° 3.78 (859).
	N-Methylpyrrole, S 1.94 (393).
C_5H_7NO	α,γ-Dimethylisoxazole, B 3.18 (704).
C_5H_8	Cyclopentene, 25° Hx 0.97, CT 0.92 (444).
	1-Methyl-1,3-butadiene, B 25° 0.50 (292); trans-, 389-469°K 0.68 (781).
	2-Methyl-1,3-butadiene, Hx -75 to 25° 0.15 (291); Hp or Am 25° 0 to 0.3 (320); 358-477°K 0.38 (781).
	n-Propylacetylene, 298°K 0.875, 348° 0.847, 398° 0.846 (596).
$C_5H_8Br_4$	Pentaerythritol tetrabromide, B 0 (35); B 25° 0 (61); Vap. 0 (287).
$C_5H_8Cl_2$	Cyclopentylidene chloride, B 25° 2.35 (841).
$C_5H_8Cl_4$	Pentaerythritol tetrachloride, B 25°< 0.2 (23); B 0 (35).
$C_5H_8I_4$	Pentaerythritol tetraiodide, B 0 (35).
$C_5H_8N_2O$	Methylethylfurazan, S 4.061 (433).
$C_5H_8N_2O_2$	Methylethylfurazan peroxide, B 25° 4.75 (666).
C_5H_8O	Cyclopentanone, S 3.00 ± .03 (108).
	Cyclopropyl methyl ketone, B 25° 2.84 (859).
	Tiglaldehyde, B 25° 3.39 (858).
$C_5H_8O_2$	Acetylacetone, B 16° 2.78 ± .08 (147); 322-477°K 3.00 (332).
	2,6-Dioxa-4-spiroheptane, S 0.79 (354).
$C_5H_8S_2$	2,6-Dithia-4-spiroheptane, S 1.12 (354).

C_5H_9Br	Cyclopentyl bromide, B 25° 2.20 (841).
$C_5H_9BrO_2$	2-Bromomethyl-1,3-dioxane, B 25° 2.89 (560).
C_5H_9Cl	Cyclopentyl chloride, B 25° 2.08 (841).
C_5H_9ClO	Isovaleryl chloride, B 20° 2.63 (498).
	n-Valeryl chloride, B 20° 2.61 (498).
C_5H_9F	Cyclopentyl fluoride, B 25° 1.86 (841).
C_5H_9I	Cyclopentyl iodide, B 25° 2.06 (841).
C_5H_9N	n-Valeronitrile, B 20° 3.57 (411); 423-522°K 4.09 (417).
	tert-Butyl cyanide, B 25° 3.65 (857).
C_5H_9NO	γ-Aminovaleric acid anhydride, B 25° 2.62 (330).
$C_5H_9NO_2$	Methyl 1-nitrosoisopropyl ketone, see $C_{10}H_{18}N_2O_4$.
C_5H_{10}	Cyclopentane, Lqd. 20° 0 (749).
	Ethylcyclopropane, Lqd. 20° 0 (749).
	2-Methyl-1-butene, Lqd. 20° 0.54 (749).
	1-n-Pentene (α-n-amylene), B 0.37 (355); Lqd. 20° 0.47 (749).
$C_5H_{10}Br_2$	1,2-Dibromo-n-pentane, B 25° 1.75 (386).
	1,5-Dibromo-n-pentane, 233°K 2.28, 253° 2.34, 273° 2.39, 293° 2.43, 313° 2.47, 333° 2.48 (184); B 25° 2.25, 50° 2.28 (259).
	2,3-Dibromo-n-pentane, B 25° 2.12 (386); dl-threo-, Lqd. 25° 1.90 (710); dl-erythro-, Lqd. 25° 1.67 (710).
$C_5H_{10}N_2O_2$	Methylethylglyoxime, D 20° 1.106 (366).
$C_5H_{10}O$	Diethyl ketone, B 15° 2.72 (66); B, T, Hx or CD 20° 2.70 (827).
	Isovaleraldehyde, B 20° 2.60 (561, 583).
	Methyl propyl ketone, B 15° 2.72 (66); B 22° 2.70 ± .03 (203).
	Tetrahydropyran (pentamethylene oxide), B 25° 1.87 (333).
	n-Valeraldehyde, B 20° 2.57 (561, 583).
$C_5H_{10}O_2$	2,2-Dimethyl-1,3-dioxolane, B 25° 1.12 (560).
	2,4-Dimethyl-1,3-dioxolane, B 25° 1.32 (560).
	Ethyl propionate, B 25° 1.81, 50° 1.78 (185); B 22° 1.74_2 ± .03 (203); B 25 and 50° 1.81 (315).
	Isobutyl formate, B 22° 1.88_0 ± .01 (203).
	Isopropyl acetate, B 22° 1.85_1 ± .03 (161).
	Isovaleric acid, B 22° 0.89 (105); B 25° 0.63 ± .01 (147).
	Methyl butyrate, B 22° 1.70_7 ± .02 (203).
	2-Methyl-1,3-dioxane, B 25° 1.89 (560).
	n-Propyl acetate, B 25° 1.86 (127); 22° B 1.78_2 ± .03, Hp 1.78_2 ± .03, CT 1.90_5 ± .03 (203).
	Trimethylacetic acid, B 30° 1.9 (699); B, Hx or D 0-40° 1.175-1.79 (798).
$C_5H_{10}O_3$	Diethyl carbonate, 353-477°K 1.06 (487, 556); B 25° 1.06 (668); B 10° 0.94, 25° 0.98, 50° 1.04, T -40° 0.67, -20° 0.72, 0° 0.77, 25° 0.84, 50° 0.93, 70° 0.97, Hx -25° 0.68, 0° 0.75, 25° 0.81, 50° 0.91 (675).
	2-Methoxyethanol acetate, B 30° 2.13 (780).
$C_5H_{10}S$	3,3-Dimethyl-1-thiacyclobutane, S 1.76 (354).
$C_5H_{11}Br$	n-Amyl bromide, B 20° 1.95 (528); B 18° 2.02 (646).
	tert-Amyl bromide, B 20° 2.25 (175).
	Isoamyl bromide, B 20° 1.93 (175).
$C_5H_{11}Cl$	n-Amyl chloride, 351-381°K 2.12 ± .01 (536).
	tert-Amyl chloride, B 20° 2.14 (175).
	Isoamyl chloride, B 20° 1.92 (175).
$C_5H_{11}F$	n-Amyl fluoride, B 25° 1.85 (857); CT 20° 1.30 (751).
	tert-Amyl fluoride, B 25° 1.92 (857); CT 20° 1.54 (751).
$C_5H_{11}I$	n-Amyl iodide, B 20° 1.88 (528); CT 20° 1.91 (750).
	tert-Amyl iodide, B 20° 2.18 (175); CT 20° 2.19 (750).
	2-Iodopentane, CT 20° 2.09 (750).

	3-Iodopentane, CT 20° $\underline{2.09}$ (750).
	Isoamyl iodide, B 20° 1.83 (175).
$C_5H_{11}N$	Piperidine, B 10-40° $\underline{1.17}$ ± .02 (382).
	Propylidene-ethylamine, B 10-40° $\underline{1.51}$ ± .01 (382).
$C_5H_{11}NO$	Valeramide, B $\underline{3.7}$ (516).
$C_5H_{11}NO_2$	n-Amyl nitrite, B 25° $\underline{2.27}$ (265); CT 25° $\underline{2.22}$ (334, 336, 390).
	Ethyl α-aminopropionate, B 25° $\underline{2.09}$ (330).
	Ethyl β-aminopropionate, B 5-75° $\underline{2.14}$ (330).
	Betaine, benzene-eth. alc. 25° 6.11 (414).
C_5H_{12}	n-Pentane, Lqd. -90 to 30° ca. $\underline{0}$ (110); Lqd. 20° $\underline{0}$ (749).
$C_5H_{12}N_2S$	sym-Diethylthiourea, D 20° 4.9 (302).
$C_5H_{12}O$	n-Amyl alcohol, Lqd. 18° $\underline{0.89}$ (4); B or Hx $\underline{1.66}$ (237); E 20°
	$\underline{1.8}$ (298, 358); B $\underline{1.66}$, E $\underline{1.8}$ (357); B 30° $\underline{1.65}$ (430); B $\underline{1.36}$
	(564); B $\underline{1.67}$ (655).
	tert-Amyl alcohol, B 24° $\underline{1.83}$ (13); B 18° $\underline{1.66}$ (234); B 30°
	$\underline{1.65}$ (430); Lqd. 30-40° $\underline{1.94}$ (816).
	Isoamyl alcohol, B 18-64° $\underline{1.76}$ (13); CT 25° $\underline{1.85}$ (24); B $\underline{1.65}$
	(54); B 10-70° $\underline{1.62}$ (55); B 25° $\underline{1.82}$ (80); B 30° $\underline{1.64}$ (430).
$C_5H_{12}O_2$	Diethoxymethane, B 25° $\underline{0.92}$ (437); 329°K $\underline{1.22}$, 352° $\underline{1.24}$,
	409-476° $\underline{1.26}$ (487, 556).
$C_5H_{12}O_3$	2-Hydroxymethyl-2-methyl-1,3-propanediol (pentaglycerol),
	D 15° $\underline{2.76}$ ± .08, 30° $\underline{2.76}$ ± .10 (707).
$C_5H_{12}O_4$	Pentaerythritol, Vap. ca. $\underline{2}$ (73, 287).
	Tetramethyl orthocarbonate, B $>\underline{0}$ (35); 374-456°K ca. $\underline{0}$ (114).
$C_5H_{12}S$	n-Amyl mercaptan, B 25° $\underline{1.50}$, 50° $\underline{1.51}$ (326).
$C_5H_{12}S_4$	Tetramethyl orthothiocarbonate, Lqd. 70° $\underline{0.50}$ (778).
$C_5H_{14}N_2$	Pentamethylenediamine, B 25° $\underline{1.91}$, 45° $\underline{1.92}$ (511, 669).
$C_6Br_3N_3O_6$	1,3,5-Tribromo-2,4,6-trinitrobenzene, B 20° and 59° $\underline{0}$ (172);
	B 20° $\underline{0.66}$ (192).
$C_6Cl_3N_3O_6$	1,3,5-Trichloro-2,4,6-trinitrobenzene, B 20° $\underline{0.64}$ (192).
$C_6Cl_4O_2$	Tetrachloro-p-benzoquinone (chloranil), S $\underline{0.38}$ (354); B $\underline{0.6}$
	(598); D 25° $\underline{0.86}$ (661).
C_6Cl_6	Hexachlorobenzene, B 50° $\underline{0.2}$ (192).
C_6HCl_5	Pentachlorobenzene, B 20° $\underline{0.87}$ ± .03 (682); B 25° $\underline{0.88}$ (701).
$C_6H_2Br_2O_2$	2,5-Dibromo-1,4-benzoquinone, B 25° $\underline{0.70}$ (418).
$C_6H_2Br_4$	1,2,4,6-Tetrabromobenzene, B $\underline{0.7}$ (168).
$C_6H_2Cl_2O_2$	2,5-Dichloro-1,4-benzoquinone, B 25° $\underline{0.64}$ (418).
$C_6H_2Cl_4$	1,2,3,4-Tetrachlorobenzene, B 25° 1.90 (701).
	1,2,4,6-Tetrachlorobenzene, B $\underline{0.65}$ (168).
$C_6H_2N_4O_6$	1,3-Dinitro-o-benzoquinone dioxime peroxide, D 25° $\underline{2.98}$ (694).
	2,3-Dinitro-o-benzoquinone dioxime peroxide, D 25° $\underline{2.74}$ (694).
	2,3-Dinitro-o-benzoquinone furoxan, D 25° $\underline{2.74}$ (741).
$C_6H_3BrN_2O_4$	1-Bromo-2,4-dinitrobenzene, B 20° $\underline{3.1}$ ± .1 (244).
	1-Bromo-3,5-dinitrobenzene, B 20° $\underline{2.3}$ (244).
$C_6H_3Br_3$	1,3,5-Tribromobenzene, B $\underline{0.2}$ (79); B 20° $\underline{0.28}$ (192).
$C_6H_3Br_3O$	2,4,6-Tribromophenol, B $\underline{1.56}$ (168).
$C_6H_3ClN_2O$	3-Chloro-o-benzoquinone furazan, D 25° $\underline{3.18}$ (581).
$C_6H_3ClN_2O_2$	3-Chloro-o-benzoquinone furoxan, D 25° $\underline{3.90}$ (741).
$C_6H_3ClN_2O_4$	1-Chloro-2,4-dinitrobenzene, B $\underline{3.29}$ (168); B 20° $\underline{3.0}$ ± .1 (244).
$C_6H_3Cl_2NO_2$	2,3-Dichloronitrobenzene, B 25° $\underline{3.86}$ (807).
	2,4-Dichloronitrobenzene, B 25° $\underline{2.66}$ (807).
	2,5-Dichloronitrobenzene, B $\underline{3.45}$ (168); B 25° $\underline{3.45}$ (807).
	2,6-Dichloronitrobenzene, B 25° $\underline{4.18}$ (807).
	3,4-Dichloronitrobenzene, B 25° $\underline{2.17}$ (807).
	3,5-Dichloronitrobenzene, B 25° $\underline{2.66}$ (807).
$C_6H_3Cl_3$	1,2,3-Trichlorobenzene, B 20° $\underline{2.31}$ ± .01 (682).
	1,2,4-Trichlorobenzene, B $\underline{1.25}$ (168); S $\underline{1.25}$ (657).

	1,3,5-Trichlorobenzene, B 20° 0.28 (192); S 0 (657).
$C_6H_3Cl_3O$	2,4,6-Trichlorophenol, B 1.62 (168).
$C_6H_3IN_2O_4$	1-Iodo-2,4-dinitrobenzene, B 20° 3.4 ± .1 (244).
$C_6H_3I_3$	1,3,5-Triiodobenzene, B 20° 0.24 (192).
$C_6H_3N_3O_3$	1-Nitro-o-benzoquinone furazan, D 25° 5.76 (694, 741).
$C_6H_3N_3O_4$	1-Nitro-o-benzoquinone furoxan (1-nitro-o-quinone dioxime peroxide), D 25° 5.47 (694, 741).

3-Nitro-o-benzoquinone furoxan (3-nitro-o-quinone dioxime peroxide), D 25° 2.50 (694, 741).

$C_6H_3N_3O_6$ 1,3,5-Trinitrobenzene, B 0.3 (17); B 25° 0.8 (58, 61); B 25° 1.08 (64); B 0.7 (79); B 10-50° 0.8 (84); B 20° 0.28 (192); N 85° 0.78 (340); N 85° 0.56 (341); B 20° 0.41 ± .07 (342); B 25 and 45° ca. 0.5, D 25° 0.4 (426); Cf 25° 0.8 (479).

C_6H_4BrCl o-Chlorobromobenzene, B 20° 2.13 (102); B 2.21, Hx 2.23 (219).
m-Chlorobromobenzene, B 1.51, Hx 1.53 (219).
p-Chlorobromobenzene, B 25° 0.1 (58); B 25°< 0.2 (61); B < 0.04 (219).

$C_6H_4BrClO_2$ p-Bromobenzenesulfonyl chloride (p-bromophenylsulfonyl chloride), B 20° 3.23 (501); B 25° 3.30 (848).

C_6H_4BrF o-Fluorobromobenzene, B 22° 2.27 (102).
p-Fluorobromobenzene, Vap. 0.5 (757).

C_6H_4BrI o-Bromoiodobenzene, B 20° 1.73 (92); B 19° 1.86 (102).
m-Bromoiodobenzene, B 20° 1.14 (92).
p-Bromoiodobenzene, B 20° 0.49 (92).

C_6H_4BrNO p-Nitrosobromobenzene, 1.92 (743).

$C_6H_4BrNO_2$ o-Bromonitrobenzene, B 25 and 65° 3.98 ± .06 (173); B 20° 4.20 (192).
m-Bromonitrobenzene, B 25 and 65° 3.17 ± .08 (173); B 20° 3.41 (192).
p-Bromonitrobenzene, B 2.69 (17); B 2.53 (79); B 25 and 65° 2.45 ± .1 (173); B 20° 2.65 (192).

$C_6H_4BrN_3$ p-Bromophenyl azide, B 23° 0.64 (156, 212).

$C_6H_4Br_2$ o-Dibromobenzene, B 1.52 (15); B 1.56 (16); S 1.67 (39); B 20° 1.87 (92); B 23° 2.03 (102); B 20° 2.11 (192); B 25° 2.12 (507); B 22° 2.26 (640); S 2.0 (657).
m-Dibromobenzene, B 1.09 (16); S 1.22 (39); B 20° 1.55 (92); B 20° 1.46 (192); S 1.5 (657).
p-Dibromobenzene, B 1.52 (15); B 0 (16); B 20° 0 (92); B 20° 0.22 (192); S 0 (657).

$C_6H_4Br_3N$ 2,4,6-Tribromoaniline, B 1.80 (168).

$C_6H_4Br_4$ 1,2,3,4-Tetrabromo-2,3-dimethylbutane, B 25° 1.55 (845).

C_6H_4ClF o-Fluorochlorobenzene, B 18° 2.33 (102).

C_6H_4ClI o-Chloroiodobenzene, B 19° 1.93 (102).
m-Chloroiodobenzene, B 25° 1.39 (643).
p-Chloroiodobenzene, B 25° 0.46 (643).

$C_6H_4ClIO_2S$ p-Iodobenzenesulfonyl chloride (p-iodo-phenylsulfonyl chloride), B 25° 3.53 (848).

C_6H_4ClNO p-Nitrosochlorobenzene, B 25° 1.80 (231); B 1.84 (678).

$C_6H_4ClNO_2$ o-Chloronitrobenzene, B 4.25 (17, 79); B 20° 3.78 (92); B 25 and 62° 3.96 ± .06 (173); B 20° 4.33 (192); B 20° 4.82 (421); 476°K 4.59 (538); Lqd. 33-160° 6.17 (543).
m-Chloronitrobenzene, B 3.38 (17, 79); B 20° 3.18 (92); B 25 and 65° 3.12 ± .08 (173); B 20° 3.40 (192); 483°K 3.69 (538); Lqd. 46-160° 4.26 (543); Lqd. 55-65° 3.31 (815).
p-Chloronitrobenzene, B 2.52 (17); B 2.55 (79); B 20° 2.36 (92); B 25 and 65° 2.34 ± .1 (173); B 20° 2.57 (192); B 20° 3.12 (421); 483°K 2.78 (538); Lqd. 84-160° 2.80 (543).

$C_6H_4ClNO_4S$ m-Nitrobenzenesulfonyl chloride (m-nitrophenylsulfonyl chloride), B 4.12 (718, 719); B 25° 4.09 (848).

$C_6H_4ClN_3$ p-Chlorophenyl azide, B 21° 0.47 (156, 212); B 25° 0.35 (189); B 25° 0.33 (316).

$C_6H_4Cl_2$ o-Dichlorobenzene, B 2 (15); B 2.0 (16); B or Hx 0-50° 2.30 (27); B 2.24 (39); B or Hp 25° 2.25 (52); B 19° 2.24 (102); B 20° 2.33 (192); B 24° 2.25 (497); 420-448°K 2.16 (538); B 22° 2.26 (640); E 20° 2.00 (723); Vap. 2.51 (757).

m-Dichlorobenzene, B 1.21 (15, 16); B or Hx 0-50° 1.55 (27); B 1.37 (39); B or Hp 25° 1.48 (52); B 20° 1.48 (192); B 24° 1.37 (497); 413 and 458°K 1.67 (538); E 20° 1.33 (723).

p-Dichlorobenzene, B 0 (15, 16); B or Hx 0-50° 0 (27); B or Hp 25° 0 (52); B 0 (60); B 20° 0.23 (192); Hp 20° 0 (220); 25° B 0, CT 0, NB 1.65, CB 0.82, benzonitrile 1.60, ethyl benzoate 0.85, dimethylaniline 0.70 (491); B 24° 0 (497); 434°K 0 (538).

$C_6H_4Cl_2O$ 2,4-Dichlorophenol, B 25° 1.59 (567).

$C_6H_4Cl_2O_2S$ p-Chlorobenzenesulfonyl chloride (p-chloro-phenylsulfonyl chloride), B 25° 3.20 (848).

$C_6H_4Cl_3I$ o-Chlorophenyl iododichloride, B 25° 2.95 (643).
m-Chlorophenyl iododichloride, B 25° 2.11 (643).
p-Chlorophenyl iododichloride, B 25° 1.3 (643).

$C_6H_4Cl_3N$ 2,4,6-Trichloroaniline, B 1.94 (168).

C_6H_4FI o-Fluoroiodobenzene, B 22° 2.00 (102).
p-Fluoroiodobenzene, Vap. 0.9 (757).

$C_6H_4FNO_2$ p-Fluoronitrobenzene, B 21° 2.63 (103); Vap. 2.87 (757).

$C_6H_4F_2$ o-Difluorobenzene, B 22° 2.38 (102).

C_6H_4INO p-Nitrosoiodobenzene, 2.16 (743).

$C_6H_4INO_2$ o-Iodonitrobenzene, B 25 and 65° 3.66 ± .07 (173); B 22° 3.92 (313).
m-Iodonitrobenzene, B 25 and 65° 3.22 ± .09 (173); B 22° 3.43 (313).
p-Iodonitrobenzene, B 25 and 65° 2.63 ± .12 (173); B 22° 3.04$_5$ (313).

$C_6H_4I_2$ o-Diiodobenzene, S 1.32 (15); B 1.82 (16); S 1.63 (39); B 23° 1.70 (102); B 20° 1.69 (192).
m-Diiodobenzene, B 1.01 (15, 16, 39); B 20° 1.27 (192).
p-Diiodobenzene, B 20° 0.19 (192).

$C_6H_4N_2$ Isonicotinonitrile, B 25° 1.61 (814).

$C_6H_4N_2O$ o-Benzoquinone furazan (o-quinone dioxime furazan), D 25° 4.37 (607, 741).

$C_6H_4N_2O_2$ o-Benzoquinone furoxan (o-quinone dioxime peroxide), D 25° 5.29 (607, 741).

$C_6H_4N_2O_2S_2$ Dithionyl-p-phenylenediamine, B 25° 1.60 (728).

$C_6H_4N_2O_3$ p-Nitronitrosobenzene, B 0.84 (678).

$C_6H_4N_2O_3S$ o-Nitrothionylaniline, B 25° 3.71 (728).
m-Nitrothionylaniline, B 25° 3.47 (728).
p-Nitrothionylaniline, B 25° 3.06 (728).

$C_6H_4N_2O_4$ o-Dinitrobenzene, B 5.95 (17); B 25° 6.05 (64); B 6.00 (79); B 20° 5.98 (192); S 6 (657).
m-Dinitrobenzene, B 4.02 (17); B 25° 3.81 (64); B 3.70 (79); B 20° 3.78 (192); B 25° 4.00 (306); N 85° 3.72 (340); N 85° 3.73$_5$ (341); S 3.79 (657); 25° B 3.96, Hx 4.07, E 3.22 (723).
p-Dinitrobenzene, B 0.3 (17); B < 0.3 (60); B 25° 0.32 (64); B 0.8 (79); B 20° 0.58 (192); B 25 and 45° ca. 0.5, D 25° 0.2 (426).

$C_6H_4N_4O_2$ p-Nitrophenyl azide, B 18° 2.96 (212).

$C_6H_4N_4O_6$ 2,4,6-Trinitroaniline, D 25° 3.25 (747).

$C_6H_4O_2$ p-Benzoquinone (quinone), B $\underline{0.67}$ (119); Vap. $\underline{0}$ (287); B 25° $\underline{0.65}$, Hx 40° $\underline{0.66}$, CT 25° $\underline{0.68}$ (418); B or CT 25 and 45° $\underline{0.69}$ (427).

$C_6H_4S_2$ trans-Thiophthene, Hp $\underline{0}$ (410).

$C_6H_4Se_2$ Isoselenophthene, B 25° ca. $\underline{0}$ (665, 670).
cis-Selenophthene, B 25° $\underline{1.52}$ (665, 670).
trans-Selenophthene, B 25° $\underline{1.07}$ (665, 670).

C_6H_5Br Bromobenzene, B $\underline{1.56}$ (17, 79); B 25° $\underline{1.51}$ (58); B 25° $\underline{1.5}$ (61); B $\underline{1.7}$ (71); B 20° $\underline{1.49}$ (102); CT 10-60° $\underline{1.53}$ (106); B 25° $\underline{1.5}_2$ (127); B 20° $\underline{1.53}$ (192); Hx -60 to 60° $\underline{1.35}$ (235); S $\underline{1.49}$ (249); B 25° $\underline{1.55}$ (372); 374-483°K $\underline{1.71}$ (417); Hx 25° $\underline{1.60}$ (453); B 25° $\underline{1.55}$ (492); B 10-40° $\underline{1.53}$, Hx -23 to 40° $\underline{1.55}$ to $\underline{1.59}$, CH 10-40° $\underline{1.56}$, T -23° $\underline{1.47}$, 0° $\underline{1.50}$, 20° $\underline{1.49}$, 40° $\underline{1.48}$, CT 0-40° $\underline{1.51}$, CD -23 to 20° $\underline{1.38}$ to $\underline{1.40}$ (527); B $\underline{1.48} \pm .05$ (568); D 30° $\underline{1.58}$ (677); CT 20° $\underline{1.68}$ (711); Vap. $\underline{1.77}$ (757).

C_6H_5BrO o-Bromophenol, B 25° $\underline{1.36}$ (145); S $\underline{1.36}$ (657); 25° D $\underline{2.36}$, CT $\underline{1.15}$ (777).

p-Bromophenol, B 25° $\underline{2.86}$ (145); B $\underline{2.12}_5$ (224); S $\underline{2.13}$ (657); 25° B $\underline{2.25}$, D $\underline{2.78}$ (777).

C_6H_5Cl Chlorobenzene, B $\underline{1.58}$ (17); B or Hx 0-50° $\underline{1.61}$ (27); B 25° $\underline{1.55}$ (31); B 18° $\underline{1.55}$ (47); B or Hp 25° $\underline{1.56}$ (52); Lqd. or Hx -80 to 80° $\underline{1.52}$ (51); 25° Hx $\underline{1.55}$, CD $\underline{1.52}$ (63); B $\underline{1.64}$ (79); B 21° $\underline{1.56}$ (102); CT 10-60° $\underline{1.59}$ (106); B 19° $\underline{1.56}$, T 19° $\underline{1.60}$ (116, 121); B $\underline{1.56}$, T $\underline{1.60}$, Cf $\underline{1.51}$ (122); B $\underline{1.61}$ (125); B 25° $\underline{1.5}_7$ (127); B or D 25° $\underline{1.5}$ (144); B 20° $\underline{1.56}$ (192); $\underline{1.56}$ (249); E 25° $\underline{1.3}$ (298, 358); N 85° $\underline{1.52}$ (340); N 85° $\underline{1.51}_8$ (341); 360-495°K $\underline{1.69} \pm .01$ (350); B $\underline{1.54}$, Hx $\underline{1.62}$, CD $\underline{1.46}$, E $\underline{1.3}$ (357); B $\underline{1.53}$ (369); B 32° $\underline{1.58}$ (416); 374-518°K $\underline{1.70}$ (432); 25° Vap. $\underline{1.69}$, Lqd. $\underline{1.22}$, B $\underline{1.59}$, Cf $\underline{1.18}$, NB $\underline{1.62}$ (493); Hp 0-60° $\underline{1.61}$ to $\underline{1.63}$, T 0-60° $\underline{1.55}$ to $\underline{1.57}$, CT 0-60° $\underline{1.55}$ to $\underline{1.58}$, CD 0-40° $\underline{1.48}$ (504); CT 20° $\underline{1.65}$ (534); Hx 20° $\underline{1.59}$ (557); Vap. $\underline{1.69}$ (637); B 22° $\underline{1.56}$ (640); D 30° $\underline{1.61}$ (677); B 20° $\underline{1.54}_5 \pm .01$ (682); CT 20° $\underline{1.64}$ (711); 25° B $\underline{1.57}$, D $\underline{1.62}$ (754); Vap. $\underline{1.72}$ (757); B 30° $\underline{1.53}$ (779); B 25° $\underline{1.60}$ (844).

$C_6H_5ClN_2O_2$ 4-Chloro-2-nitroaniline, B 25° $\underline{4.41}$ (747).

C_6H_5ClO o-Chlorophenol, B 25° $\underline{1.3}$ (61); B 25° $\underline{1.43}$ (145); B $\underline{1.30}_9$ (224); Hx 10-40° $\underline{1.11}$, Hp -30 to 70° $\underline{1.10}$, CT -15 to 30° $\underline{1.14}$ to $\underline{1.16}$ (546); S $\underline{1.31}$ (657); 423-598°K $\underline{2.17} \pm .04$ (688); 25° B $\underline{1.33}$, D $\underline{2.11}$, CT $\underline{1.15}$ (777); Lqd. 30-40° $\underline{1.55}$ (816).

m-Chlorophenol, B 25° $\underline{2.17}$ (145); B $\underline{2.10}_1$ (224); S $\underline{2.10}$ (657).

p-Chlorophenol, B 25° $\underline{2.4}$ (61); B 25° $\underline{2.68}$ (145); B $\underline{2.22}_2$ (224); S $\underline{2.22}$ (657); Lqd. 55-65° $\underline{2.10}$ (816).

$C_6H_5ClO_2S$ Benzenesulfonyl chloride, B 20° $\underline{4.47}$ (501); B $\underline{4.54}$ (718, 719); B 25° $\underline{4.53}$ (848).

Phenyl chlorosulfite, B 25° $\underline{2.43}$, 45° $\underline{2.44}$ (456).

$C_6H_5Cl_2I$ Phenyl iododichloride, B 25° $\underline{2.61}$ (643).

$C_6H_5Cl_2N$ 2,5-Dichloroaniline (p-dichloroaniline), B $\underline{1.68}$ (168).

C_6H_5F Fluorobenzene, B 20° $\underline{1.39}$ (92); B 21° $\underline{1.45}$ (102); S $\underline{1.45}$ (249); B 25° $\underline{1.41}$ (334, 390); B 30° $\underline{1.46}$ (343); 344-507°K $\underline{1.57}$ (432); B 25° $\underline{1.47}$ (492); CT 20° $\underline{1.44}$ (711); Vap. $\underline{1.57}$ (757); D 30° $\underline{1.50}$ (782).

C_6H_5FO o-Fluorophenol, 25° D $\underline{1.84}$, CT $\underline{1.16}$ (777).

C_6H_5I Iodobenzene, B 20° $\underline{1.25}$ (92); B 20° $\underline{1.295}$ (102); B 20° $\underline{1.38}$ (192); Hx -60 to 60° $\underline{1.25}$ (235); CT 20° $\underline{1.69}$ (751).

$C_6H_5IN_2O_2$ 4-Iodo-2-nitroaniline, B 25° $\underline{4.55}$ (747).

C_6H_5NO Nitrosobenzene, B $\underline{3.22}$ (119); B 25° $\underline{3.14}$ (231); B $\underline{3.085}$ (678).

C_6H_5NOS Thionylaniline, B 20° $\underline{2.6}$ (215); B 25° $\underline{1.90}$ (728).

$C_6H_5NO_2$	Isonicotinic acid, D 25° 2.7 (814).

Nitrobenzene, Lqd. 18° 0.71 (4); B 24-65°, T 24-100°, CD 24° 3.84 (13); 3.7 (39); Hx or CD 25° 3.89 (63); B 25° 3.90 (64); B 21° 4.08 (103); B 19° 3.98, T 19° 3.83, Cf 18° 3.30, CB 19° 2.83 (116, 121); B 3.98, T 3.83, Cf 3.24, CB 2.84 (122); B 27° 3.916, CT 10-50° 3.92 to 3.96 (129); B 20° 3.97 (192); 3.96 (249); S 3.79 (250); Hx 25° 4.0₀ (251); B 25° 3.82 (278); E 25° 3.2 (298, 358); 326-480°K 0 (308); Hx 25° 3.96 (312); B 22° 3.93 (313); CT 25° 3.97 (334, 390); N 85° 3.73 (340); N 85° 3.70 (341); pX or decane 20-40° 4.28, 60-120° 4.22 (345); 402-523°K 4.23 ± .01 (350); B 3.96, Hx 4.05, CD 3.74, E 3.2 (357); 25° B 3.94, Hx 4.05, CH 3.97, CT 3.93, CD 3.66, Cf 3.17, Dec. 3.93 (360); 25° B 3.936, Hx 4.049, CH 3.974, CT 3.932, CD 3.658, Cf 3.172, Dec. 3.930 (361); D 25° 3.93, 143° 4.04 (362); B 4.03 (369); 10-40° B 3.96 to 3.99, Hx 4.08 to 4.10, Hp 4.10, CH 4.06 to 4.09, D 3.90 to 3.94, CT 4.00 to 4.02 (381); S 4.4 (392); B 32° 3.97 (416); B 25° 3.95 (428); 442-549°K 4.19 (432); B 25° 4.03 (492); 298°K 4.23, 25° Lqd. 1.72, B 4.05, CB 2.49, Cf 3.05 (493); 25° B 4.01, D 4.03 (747) 30° B 3.96, D 3.99 (779).

p-Nitrosophenol, D 20° 4.72 (280). |
| $C_6H_5NO_3$ | o-Nitrophenol, B 25° 3.10 (145); B 25° 3.11 (480); Lqd. 50-60° 3.01 (816).

m-Nitrophenol, B 25° 3.90 (145).

p-Nitrophenol, B 25° 5.05 (145); B 5.01₆ (224). |
| $C_6H_5N_3$ | 1,2,3-Benzotriazole, D 25° 4.07 (787).

Phenyl azide, B 22° 1.55 (156, 212); B 25° 1.55 (189, 316). |
| $C_6H_5N_3O_4$ | 2,3-Dinitroaniline, D 25° 7.30 (747).

2,4-Dinitroaniline, D 25° 6.48 (747).

2,5-Dinitroaniline, D 25° 2.67 (747).

2,6-Dinitroaniline, B 25° 1.88 (747).

3,4-Dinitroaniline, D 25° 8.90 (747).

3,5-Dinitroaniline, D 25° 5.91 (747). |
| C_6H_6 | Benzene, Lqd. 0 (8); 278-353°K 0 or 0.2 (19); Cf 25° 1.15 (24); 25° CD 0.09, Hx 0.08 (63); Cf 0.53 (116, 121, 122); Hp or CT 20° 0 (220); 293-333°K 0 (295); E 25° 0 (298, 358); 326-480°K 0 (308); 346-522°K 0 (350); 25° B 0, Cf 0, CB 0.73, NB 1.51 (491); Hx 20° 0 (557); B 0 (564); Lqd. 20 and 50° 0.14 (720). |
| C_6H_6BrN | o-Bromoaniline, B 20° 1.77 (193).

m-Bromoaniline, B 20° 2.65 (193).

p-Bromoaniline, B 20° 2.99 (193); B 23° 2.87 (215); B 25ᵁ 2.85 (492). |
| $C_6H_6BrNO_2S$ | p-Bromobenzenesulfonamide, D 25° 4.41 (849). |
| C_6H_6ClN | o-Chloroaniline, B 25° 1.84 (163); B 1.81 (168); B 20° 1.77 (193); S 1.79 (657).

m-Chloroaniline, B 25° 2.91 (163); B 2.70 (168); B 20° 2.66 (193); S 2.68 (657).

p-Chloroaniline, B 18° 2.93 (111); B 18° 2.93 (116); B 25° 3.00 (163); B 20° 2.97 (193); B 24° 2.90 (215); S 2.92 (657). |
$C_6H_6ClNO_2S$	p-Chlorobenzenesulfonamide, 25° B 3.91, D 4.39 (849).
$C_6H_6Cl_6$	1,2,3,4,5,6-Hexachlorocyclohexane (benzene hexachloride), cis-, B 2.20 (118); B 25° 2.15 (201); 2.2 (838); trans-, B 0 (118); B 25° 0.70 (201); B 60° 0 (232); 0 (838); γ-, 3.6 (838); δ-, 0 (838).
C_6H_6FN	p-Fluoroaniline, B 24° 2.75 (215).
C_6H_6IN	p-Iodoaniline, B 25° 2.82 (215).
$C_6H_6INO_2S$	p-Iodobenzenesulfonamide, D 25° 4.50 (849).

$C_6H_6N_2O$ Isonicotinamide, D 25° 3.88 (814).

$C_6H_6N_2O_2$ o-Benzoquinone dioxime, D 25° 3.84 (607).

p-Benzoquinone dioxime, D 25° 2.37 (607).

o-Nitroaniline, B 4.45 (79); B 20° 4.25 (193); B 20° 4.01 (421); B 25° 4.26 (747); Lqd. 90-110° 5.00 (815).

m-Nitroaniline, B 4.72 (79); B 40° 4.94 (193); 30° B 5.00, D 5.22 (779).

p-Nitroaniline, B 7.1 (79); B 70° 6.4 (193); Vap. 5.6 ± 10% (287, 328); D 30° 6.68 (363); B 25° 6.32 (492); 25° B 6.17, D 6.81 (747); Lqd. 160-180° 7.16 (815).

$C_6H_6N_4O_4$ 2,4-Dinitrophenylhydrazine, B 17° 5.8 ± .6 (457).

$C_6H_6N_2O_4S$ m-Nitrobenzenesulfonamide, D 25° 4.93 (849).

C_6H_6O Phenol, B 25° 1.7 (30); B 25° 1.70 (31); CD 20° 1.63 (63); B 1.57 (224); B 70° 1.5 (431); 450°K 1.40 (538).

$C_6H_6O_2$ Catechol (pyrocatechol), 25° B 2.58, D 2.93 (809); B 27° 2.62 ± .03 (813).

Hydroquinone, E 2.47 (116); B 44° 1.4 ± .10 (813).

Resorcinol, B 44° 2.07 ± .02 (813).

$C_6H_6O_3S$ Benzenesulfonic acid, B 3.77 (718, 719).

C_6H_6S Thiophenol, B 20° 1.33 (238).

$C_6H_7BrN_2$ p-Bromophenylhydrazine, B 16° 2.89 ± .05 (457).

C_6H_7N Aniline, B 20-50° 1.6 (40); B 1.51 (79); B 18° 1.55 (121); B 20° 1.52 (193); B 1.54, Hx 1.50, E 1.67 (357); B 32° 1.52 (416); B 25° 1.51 (492); 459°K 1.48 (538); 20° B 1.53, T 1.52, Hx 1.48, CH 1.49 (585); D 25° 1.90 (732); 25° B 1.54, D 1.77 (747); B 20° 1.54 (826).

2- or α-Picoline, B 10-40° 1.72 ± .01 (382).

3- or β-Picoline, B 10-40° 2.30 ± .01 (382).

4- or γ-Picoline, B 25° 2.57 (814).

C_6H_7NO 4-Methoxypyridine, B 25° 2.94 (814).

$C_6H_7NO_2S$ Benzenesulfonamide, B 4.75 (718, 719); D 25° 5.09 (732); 25° B 4.73, D 5.09 (849).

$C_6H_7N_3O_2$ 1-Nitro-2,4-diaminobenzene, D 25° 7.11 (747).

1-Nitro-3,5-diaminobenzene, D 25° 5.86 (747).

p-Nitrophenylhydrazine, B 16° 7.2 ± .5 (457).

$C_6H_8Br_4$ 1,2,4,5-Tetrabromocyclohexane, B R.T. 2.2$_2$ (644).

$C_6H_8Cl_2O_4$ Dimethyl α,β-dichlorosuccinate, dl-, B 2.93 (118); meso-, B 2.47 (118).

$C_6H_8N_2$ 2,5-Dimethylpyrazine, B 0 (273); B 20-50° 0 (334, 335, 390).

2,6-Dimethylpyrazine, B 0.5 (273); B 20-50° 0.53 (334, 335, 390).

Hexanedinitrile, B 25° 3.54, 75° 3.66 (569); B 25° 3.76, 75° 3.87 (669).

o-Phenylenediamine, B 25° 1.45 (146); B 20-30° 1.44 (193); 423-598°K 1.48 ± .08 (688).

m-Phenylenediamine, B 25° 1.8 (146); B 20° 1.79 (193); 423 to 598°K 1.70 (688).

p-Phenylenediamine, B 25° < 0.3 (61); B 25° ca. 0 (65); B 25° 1.51 (69); B 25° 0.3 (95); B 25° ca. 1.5 (146); B 40° 1.56 (193); 423-598°K 1.46 (688).

Phenylhydrazine, B 18-19° 1.65 to 1.79 (275).

$C_6H_8N_2O_2S$ m-Aminobenzenesulfonamide (metanilamide), D 25° 5.63 (732).

p-Aminobenzenesulfonamide (sulfanilamide), D 6.60 (718, 719); D 25° 6.63 (732); D 25° 6.5 ± .1 (759); D 25° 6.62 (849).

$C_6H_8O_2$ 1,4-Cyclohexanedione, S 1.6 (117); B 25° 1.29 (427).

$C_6H_8O_6$ Ascorbic acid, 1-, D 25° 3.93 (684).

C_6H_9Br 1-Bromo-1-hexyne, B 25° 1.06 (376).

C_6H_9Cl	1-Chloro-1-hexyne, B 25° 1.23 (376).
C_6H_9I	1-Iodo-1-hexyne, B 25° 0.75 (376).
C_6H_9N	Cyclopentyl cyanide, B 25° 3.71 (841).
C_6H_9NO	α,β,γ-Trimethylisoxazole, B 3.42 (704).
$C_6H_9S_3$	Trithioacetaldehyde, α-, 2.14 (832); β-, 2.14 (832).
C_6H_{10}	n-Butylacetylene, 298°K 0.899, 348° 0.882, 398° 0.858 (596).
	Cyclohexene, 25° Hx 0.75, CT 0.63 (444); 308-480°K 0.61 (547).
	1,1-Dimethyl-1,3-butadiene, Hx -75 to 50° 0.52 (291).
	1,2-Dimethyl-1,3-butadiene, B 25° 0.53 (292); 399-487°K 0.63 (829).
	1,3-Dimethyl-1,3-butadiene, B 25° 0.59 (292); 399-497°K 0.65 (829).
	1,4-Dimethyl-1,3-butadiene, low boiling, B 25° 0.36 (292); high boiling, B 25° 0.31 (292).
	2,3-Dimethyl-1,3-butadiene, Hx -75 to 50° 0 (291); B 25° 0 (292).
	2-Ethyl-1,3-butadiene, 384-479°K 0.45 (829).
$C_6H_{10}Br_2$	1,4-Dibromocyclohexane, trans-, 0 (167); B 18° 0 (230).
	1,4-Dibromo-2,3-dimethyl-2 butene, cis-, B 25° 2.49 (845); trans-, B 25° 1.72 (845).
$C_6H_{10}Cl_2$	1,4-Dichlorocyclohexane, trans-, B 18° 0 (230).
$C_6H_{10}I_2$	1,4-Diiodocyclohexane, cis-, B 18° 2.43 (230); trans-, 0 (167); B 18° 0 (230).
$C_6H_{10}O$	Butoxyacetylene, B 25° 2.03 (796).
	Cyclohexanone, B 15° 2.75 (99); S 22° 2.75 ± .05 (108); B 25° 2.8 (143); D 25° 2.90 (756).
$C_6H_{10}O_3$	Ethyl acetoacetate, 394°K 2.95, 410° 2.92, 431° 2.95 (332); enol-, CD -80° 2.04 (518); keto-, B R.T. 3.22 (518).
$C_6H_{10}O_4$	Adipic acid, D 4.04 (570).
	Diethyl oxalate, B 25° 2.49, 50° 2.52 (185); B 25 and 50° 2.49 (315).
	1,4,5,8-Naphthodioxane, cis-, B 20° 1.90 (297); S 1.90 (354, 406); trans-, B 20° 0.79 (297); S 0.72 (354, 406).
$C_6H_{10}O_6$	Dimethyl tartrate, dl-, B 25° 2.92 (199); B 25° 2.94 (202); d-, B 25° 2.93 (199, 200).
$C_6H_{11}Br$	Bromocyclohexane, B 25° 2.3 (143); B 2.11 (232).
$C_6H_{11}BrO_2$	2-Acetoxy-3-bromobutane, dl-threo-, Lqd. 25° 2.26 (710); dl-erythro-, Lqd. 25° 2.23 (710).
$C_6H_{11}Cl$	Chlorocyclohexane, B 25° 2.3 (143); B or D 25° 2.3 (144); B 2.07 (232); B 18° 2.10 (234).
$C_6H_{11}I$	Iodocyclohexane, B 1.98 (232).
$C_6H_{11}N$	Isoamyl cyanide, B 25° 3.53 (857).
$C_6H_{11}NO$	N-Methyl-2-oxopiperidine (N-methyl-α-piperidone), B 25° 4.01 (821).
C_6H_{12}	Cyclohexane, B 25° 0.2 (88); B 25° 0 (143).
	Ethylcyclobutane, Lqd. 20° 0.05 (749).
	Methylcyclopentane, Lqd. 20° 0 (749).
	n-Propylcyclopropane, B 20° 0.75 (520).
$C_6H_{12}Br_2$	1,6-Dibromo-n-hexane, Hp 25° 2.39, 50° 2.41 (320).
	3,4-Dibromo-n-hexane, dl-, Lqd. 25° 2.06 (710); meso-, Lqd. 25° 1.57 (710).
$C_6H_{12}Cl_2$	2,3-Dimethyl-2,3-dichlorobutane, B 1.35, CT 0 (208).
$C_6H_{12}N_2$	Dimethylketazine, Vap. 1.51 (712).
$C_6H_{12}N_2O_2$	Methyl-n-propylglyoxime, D 20° 1.25 (366).
$C_6H_{12}N_2O_3$	Glycylglycine ethyl ester, B 50° 3.20 (330).
$C_6H_{12}N_4$	Hexamethylenetetramine (urotropine), Cf 25-45° nil (599).
$C_6H_{12}O$	Cyclohexanol, B 25° 1.9 (143); B 18° 1.69 (234); D 25° 1.82

(756); Lqd. 30-45° $\underline{2.83}$ (816).

Methyl n-butyl ketone, B 15° $\underline{2.73}$ (66); B 22° $\underline{2.66}_6$ ± .02 (203).

Methyl tert-butyl ketone (pinacolin), B 15° $\underline{2.79}$ (66).

Vinyl isobutyl ether, B 25° $\underline{1.20}$ (858).

Vinyl n-butyl ether, B 25° $\underline{1.25}$ (858).

$C_6H_{12}O_2$ n-Amyl formate, 376-516°K $\underline{1.90}$ (272).

n-Butyl acetate, B 25° $\underline{1.8}_5$ (127); B 22° $\underline{1.84}$ ± .01 (203); Lqd. 30-40° $\underline{1.83}$ (815).

tert-Butyl acetate, B 22° $\underline{1.91}_2$ ± .025 (161).

2,3-Dimethyl-2-butene-1,4-diol, cis-, B 25° $\underline{2.52}$ (845); trans-, B 25° $\underline{1.93}$ (845).

Isobutyl acetate, B 25° $\underline{1.8}_7$ (127); B 22° $\underline{1.85}_4$ ± .01 (203).

Ethyl n-butyrate, B 22° $\underline{1.73}_8$ ± .02 (203).

Methyl n-valerate, B 22° $\underline{1.60}_6$ ± .03 (203).

n-Propyl propionate, B 22° $\underline{1.76}_7$ ± .03 (203).

$C_6H_{12}O_3$ 2-Ethoxyethanol acetate ("Cellosolve" acetate), B 30° $\underline{2.22}$ (780); Lqd. 30-50° $\underline{2.32}$ (815).

Paraldehyde, B 18° $\underline{1.92}$ (78); B $\underline{1.93}$, T $\underline{1.99}$ (116, 122); Cf $\underline{1.99}$ (121); 25° Lqd. $\underline{1.89}$, B $\underline{2.03}$, CT $\underline{2.12}$, Cf $\underline{2.26}$, CB $\underline{1.77}$, dimethylaniline $\underline{1.68}$, benzonitrile $\underline{1.94}$, NB $\underline{1.97}$, ethyl benzoate $\underline{1.76}$ (494).

$C_6H_{13}Br$ n-Hexyl bromide, B 18° $\underline{1.97}$ (646).

$C_6H_{13}ClO_2S$ n-Hexyl chlorosulfite, B 25° $\underline{2.70}$, 45° $\underline{2.75}$ (455).

$C_6H_{13}I$ n-Hexyl iodide, CT 20° $\underline{1.92}$ (750).

$C_6H_{13}N$ Cyclohexylamine, B 25° $\underline{1.32}$ (652).

N-Methylpiperidine, B 10-40° $\underline{0.91}$ ± .01 (382).

$C_6H_{13}NO$ N-Diethylacetamide, D 30° $\underline{3.72}$ (363).

Caproamide, B $\underline{3.9}$ (516).

$C_6H_{13}NO_2$ Ethyl α-aminobutyrate, B 25° $\underline{2.13}$ (330).

Ethyl β-aminobutyrate, B 25° $\underline{2.11}$ (330).

Methyl α-aminovalerate, B 20° $\underline{1.6}$ (40).

Methyl δ-aminovalerate, B 20-50° $\underline{2.7}$ (40).

C_6H_{14} n-Hexane, Lqd. -90 to 70° $\underline{0}$ (51); 25° B $\underline{0.05}$, CD $\underline{0.08}$ (63); Lqd. -90 to 50° ca. $\underline{0}$ (110); Lqd. -60 to 60° $\underline{0}$ (235); 338-558°K $\underline{0}$ (389).

$C_6H_{14}O$ Ethyl n-butyl ether, S $\underline{1.2}$ (654); 25° B $\underline{1.24}$, kerosene $\underline{1.19}$ (836).

n-Hexyl alcohol, B 25° $\underline{1.64}$ (74, 82).

n-Propyl ether, 368-448°K $\underline{1.02}$ ± .05 (86, 254); Hx -5 to 50° $\underline{1.16}$ ± .05 (245); 331-473°K $\underline{1.18}$ ± .01 (537).

$C_6H_{14}O_2$ 1,1-Diethoxyethane (ethylidene diethyl ether), 328°K $\underline{1.07}$, 352° $\underline{1.08}$, 410° $\underline{1.08}$, 476° $\underline{1.21}$ (487, 556).

Ethylene glycol mono-n-butyl ether, B 24° $\underline{2.08}$ (633).

Hexamethylene glycol, D 25 and 50° $\underline{2.48}$ (186).

2-Methyl-2,4-pentanediol, Hp $\underline{2.1}$, D $\underline{2.9}$ (266).

$C_6H_{14}S$ n-Propyl sulfide, B 20° $\underline{1.55}$ (238).

$C_6H_{15}N$ Triethylamine, B 25° $\underline{0.90}$ (163); 298-377°K $\underline{0.82}$ (165); 25° B $\underline{0.78}_9$, Hx $\underline{0.74}_9$ (539).

$C_6H_{15}NO_3$ Triethanolamine, D 25° $\underline{3.57}$ (443).

$C_6H_{16}N_2$ Hexamethylenediamine, B 25° $\underline{1.91}$, 45° $\underline{1.94}$ (669).

$C_7H_2Br_3N_3$ 2,4,6-Tribromobenzenediazo cyanide, cis-, B 25° $\underline{2.5}$ (600); trans-, B 25° $\underline{4.0}$ (600).

$C_7H_2Cl_3N$ 2,4,6-Trichlorobenzonitrile, B $\underline{3.88}$ (168).

$C_7H_3ClN_2O_5$ 3,5-Dinitrobenzoyl chloride, B 20° $\underline{1.20}$ (499).

$C_7H_3Cl_2N$ p-Dichlorobenzonitrile, B $\underline{3.79}$ (168).

$C_7H_3Cl_5$ 2,3,4,5,6-Pentachlorotoluene, B 25° $\underline{1.55}$ (701).

C_7H_4BrClO	p-Bromobenzoyl chloride, B 20° 2.03 (499).
C_7H_4BrN	p-Bromobenzonitrile, B 21° 2.64 (103).
C_7H_4BrNS	p-Bromophenylisothiocyanate, B 20° 1.54 (215).
$C_7H_4BrN_3$	o-Bromobenzenediazo cyanide, cis-B 25° 3.79 (600); trans-, B 25° 5.32 (600).
	p-Bromobenzenediazo cyanide, cis-, B 25° 2.91 (600); trans-, B 25° 3.78 (600).
C_7H_4ClN	p-Chlorobenzoisonitrile, B 25° 2.07 (115); B 22° 2.50$_5$ (329).
	o-Chlorobenzonitrile, B 21° 4.76 (103); B 4.73 (168).
	m-Chlorobenzonitrile, B 22° 3.38 (329).
	p-Chlorobenzonitrile, S 23° 2.61 (217); B 22° 2.08$_5$ (329).
C_7H_4ClNO	p-Chlorophenyl isocyanate, CT 25° 0.84 (316).
$C_7H_4ClNO_3$	p-Nitrobenzoyl chloride, B 20° 1.11 (499).
C_7H_4ClNS	p-Chlorophenyl isothiocyanate, B 19° 1.55 (215).
	p-Chlorophenyl thiocyanate, B 21° 2.93 (103).
C_7H_4ClNSe	p-Chlorophenyl selenocyanate, B 25° 3.28 (860).
$C_7H_4ClN_3$	p-Chlorobenzenediazo cyanide, cis-, B 25° 2.93 (600); trans-, B 25° 3.73 (600).
$C_7H_4Cl_2O$	p-Chlorobenzoyl chloride, B 20° 2.00 (499).
$C_7H_4Cl_4$	p-Chlorobenzo trichloride, B 30° 0.78 (692).
C_7H_4IN	p-Iodobenzonitrile, S 23° 2.81 (217).
$C_7H_4N_2O_2$	o-Nitrobenzonitrile, B 25° 6.19 (190).
	m-Nitrobenzonitrile, B 25° 3.78 (190).
	p-Nitrobenzonitrile, B 18° 0.72 (111, 116); B 25° 0.66 (115).
$C_7H_4N_2O_2S$	p-Nitrophenyl thiocyanate, B 25° 3.10 (863).
$C_7H_4N_2O_2Se$	p-Nitrophenyl selenocyanate, B 25° 3.58 (863).
$C_7H_4N_4O_2$	p-Nitrobenzenediazo cyanide, cis-, B 25° 2.04 (600); trans-, B 25° 1.47 (600).
C_7H_5BrO	Benzoyl bromide, B 20° 3.37 (499).
	p-Bromobenzaldehyde, D 25° 2.20 (375); D 25° 2.19 (443).
$C_7H_5BrO_2$	o-Bromobenzoic acid, D 30° 2.50 (782).
	m-Bromobenzoic acid, D 30° 2.15 (677).
	p-Bromobenzoic acid, D 30° 2.08 (677).
$C_7H_5Br_3$	3,5-Dibromobenzyl bromide, B 30° 1.66 (692).
	2,4,6-Tribromotoluene, B 30° 0.73 (692).
C_7H_5ClO	Benzoyl chloride, B 3.23 (247); B 20° 3.33 (499); B 25° 3.13 (545).
	p-Chlorobenzaldehyde, B 20° 2.03 (583).
$C_7H_5ClO_2$	o-Chlorobenzoic acid, D 30° 2.43 (782).
	m-Chlorobenzoic acid, D 30° 2.20 (677).
	p-Chlorobenzoic acid, D 30° 2.00 (677).
$C_7H_5ClO_4S$	m-Sulfochlorobenzoic acid (benzoic acid-m-sulfonyl chloride), B 3.85 (718, 719); B 25° 3.84 (848).
$C_7H_5Cl_3$	α-Trichlorotoluene (benzotrichloride), B 20° 2.15 (176); B 25° 2.074 (190); CT 2.1 (208).
	2,4,6-Trichlorotoluene, B 30° 0.57 (692); B 25° 0.54 (701).
$C_7H_5FO_2$	o-Fluorobenzoic acid, D 30° 2.10 (782).
	m-Fluorobenzoic acid, D 30° 2.16 (782).
	p-Fluorobenzoic acid, D 30° 1.99 (782).
C_7H_5N	Benzoisonitrile (phenylcarbylamine, phenylisocyanide), B 18° 3.49 (111); B 18° 3.49 (116); B 22° 3.53 (252).
	Benzonitrile, B 20° 3.84 (97); B 20° 3.74, 40° 3.79, 60° 3.82 (98); B 21° 3.91 (103); B 18° 3.90 (111); B 18° 3.93, 40° 3.92, 60° 3.88 (116, 120); B 22° 3.94 (252, 329); 383-525°K 4.39 ± .02 (350); S 4.3 (392); B 20° 4.02, T -79° 3.81, -23° 3.91, 0° 3.94, 20° 3.95, Hx -23° 4.09, 20° 4.14, CH 20° 4.10, CT 0° 4.03, 40° 4.05, CD -79° 3.61, -23° 3.72, 0° 3.75, 20° 3.77 (469).

C_7H_5NO Anthranil, B 25° <u>3.06</u> (787).

Benzoxazole, B 25° <u>1.47</u> (787).

Indoxazene (<u>4,5-benzoisoxazole</u>), B 25° <u>3.03</u> (787).

Phenyl isocyanate, B 18° <u>2.34</u> (111); B 18° <u>2.34</u> (116); CT 25° <u>2.23</u> (316); B 20° <u>2.28</u> (442, 467).

Salicylonitrile, 25° B <u>4.38</u>, CT <u>3.2</u>, D <u>5.03</u> (847).

$C_7H_5NO_3$ m-Nitrobenzaldehyde, S <u>3.28</u> (117).

p-Nitrobenzaldehyde, B 25° <u>2.4</u> (58, 61); B 20° <u>2.41</u> (583).

$C_7H_5NO_4$ p-Nitrobenzoic <u>acid</u>, B 25° <u>3.5</u> (58, 61); D 25° <u>4.02</u> (460).

C_7H_5NS Benzothiazole, B 25° <u>1.45</u> (768).

Phenyl isothiocyanate, B 24° <u>2.76</u> (103); B 20° <u>3.00</u> (215).

Phenyl thiocyanate, B 23° <u>3.59</u> (103).

$C_7H_5NS_2$ 2-Mercaptobenzothiazole, B 25° <u>4.00</u> (768).

C_7H_6BrCl p-Bromobenzyl chloride, Hp 25° <u>1.71</u>, 50° <u>1.72</u> (258).

p-Chlorobenzyl bromide, Hp 25° <u>1.72</u>, 50° <u>1.74</u> (258).

$C_7H_6BrNO_2$ p-Bromophenylnitromethane, B 25° <u>2.85</u> (471).

p-Nitrobenzyl bromide, B 25° <u>3.55</u>, 50° <u>3.58</u> (258).

$C_7H_6Br_2$ 3,5-Dibromotoluene, B 30° <u>1.84</u> (692).

$C_7H_6ClNO_2$ o-Nitrobenzyl chloride, 30° B <u>4.10</u>, CT <u>3.91</u> (276); pX 20-120° <u>3.93</u> to <u>3.97</u> (289).

m-Nitrobenzyl chloride, 30° B <u>3.82</u>, CT <u>3.89</u> (276); pX 20-120° <u>3.71</u> to <u>3.80</u> (289).

p-Nitrobenzyl chloride (2-methyl-5-nitrophenylsulfonyl chloride), B 20° <u>3.63</u> (155); B 25° <u>3.55</u>, 50° <u>3.58</u> (258); pX 20-120° <u>3.45</u> (289).

$C_7H_6ClNO_4S$ p-Nitrotoluene-o-sulfonyl chloride, B <u>4.85</u> (718, 719); B 25° <u>4.82</u> (848).

$C_7H_6Cl_2$ Benzal chloride, B 20° <u>2.05</u> (176); B 25° <u>2.03</u> (190).

o-Chlorobenzyl chloride, 30° B <u>2.39</u>, Hp <u>2.25</u>, CT <u>2.30</u> (276).

m-Chlorobenzyl chloride, 30° B <u>2.05</u>, Hp <u>2.09</u>, CT <u>2.05</u> (276).

p-Chlorobenzyl chloride, B 20° <u>2.11</u> (155); 30° B <u>1.71</u>, Hp <u>1.69</u>, CT <u>1.72</u> (276); B 25° <u>1.7</u>₄ (327).

3,5-Dichlorotoluene, B 30° <u>1.88</u> (692).

$C_7H_6N_2$ Benzimidazole, D 25° <u>3.93</u> (787); D 25° <u>4.08</u> (821).

Indazole (<u>benzpyrazole</u>), B 25° <u>1.83</u> (787).

$C_7H_6N_2O$ 1-Methyl-o-benzoquinone furazan (<u>1-methyl-o-quinone dioxime furazan</u>), D 25° <u>4.50</u> (607, 741).

3-Methyl-o-benzoquinone furazan, D 25° <u>4.91</u> (607, 741).

$C_7H_6N_2O_2$ 1-Methyl-o-benzoquinone furoxan (<u>1-methyl-o-quinone furazan peroxide</u>), D 25° <u>4.93</u> (607, 741).

3-Methyl-o-benzoquinone furoxan, D 25° <u>5.38</u> (607, 741).

$C_7H_6N_2O_3$ p-Nitrobenzamide, D 20° <u>4.9</u> (284).

$C_7H_6N_2O_4$ 2,3-Dinitrotoluene, B <u>5.77</u> (850).

2,4-Dinitrotoluene, B <u>3.75</u>, CT <u>3.87</u> (850).

2,5-Dinitrotoluene, B <u>0.94</u> (850).

2,6-Dinitrotoluene, B <u>2.95</u>, CT <u>2.74</u> (850).

3,4-Dinitrotoluene, B <u>6.32</u>, CT <u>6.21</u> (850).

3,5-Dinitrotoluene, B <u>4.05</u> (850).

$C_7H_6N_2S$ p-Aminophenyl thiocyanate, B 25° <u>5.16</u> (863).

$C_7H_6N_2Se$ p-Aminophenyl selenocyanate, B 25° <u>5.22</u> (863).

C_7H_6O Benzaldehyde, B 25° <u>2.75</u> (58, 61); B 25° <u>2.77</u> (533).

$C_7H_6O_2$ Benzoic <u>acid</u>, B 25° <u>1.0</u> (30); B 22° <u>0.56</u> (105); D 25° <u>1.71</u> (460); D 30° <u>1.78</u> (677); B 30° <u>1.64</u> (699).

o-Hydroxybenzaldehyde (<u>salicylaldehyde</u>), S <u>2.90</u> (620); 25° B <u>2.88</u>, D <u>2.99</u> (809); Lqd. 30-40° <u>3.13</u> (815); Lqd. 30-40° <u>3.07</u> (816).

p-Hydroxybenzaldehyde, D 25° <u>4.18</u> (375); D 25° <u>4.19</u> (443).

o-Phenylenemethylene dioxide, S 1.0 (466).

$C_7H_6O_3$ o-Hydroxybenzoic acid (salicylic acid), D 25° 2.63 (460).
m-Hydroxybenzoic acid, D 25° 2.37 (460).
p-Hydroxybenzoic acid, D 25° 2.73 (460).

C_7H_7Br Benzyl bromide, B 25 and 50° 1.85, Hp 25 and 50° 1.87 (258);
Hx 20° 1.88 (609).
o-Bromotoluene, B 20° 1.44 (192).
m-Bromotoluene, B 20° 1.75 (192).
p-Bromotoluene, B 2.15 (71); B 20° 1.93 (192); B 25° 1.95 to
1.97 (296).

C_7H_7BrO o-Bromoanisole, 25° B 2.47, D 2.56 (777).
p-Bromoanisole, B 25° 2.23, 50° 2.25, Hp 25° 2.30, 50° 2.31
(260); Lqd. 30-40° 2.12 (815).

C_7H_7Cl Benzyl chloride, B 22° 1.84$_5$ (103); B 25° 1.8$_7$ (142); B 20°
1.85 (176); B 25° 1.82 (190); Dec 20-160° 1.72 (289).
o-Chlorotoluene, B 20° 1.39 (92); B 20° 1.43 (192); S 1.32$_4$ ±
.02 (204).
m-Chlorotoluene, B 20° 1.60 (92); B 20° 1.77 (192); S 1.79$_4$ ±
.04 (204).
p-Chlorotoluene, B 20° 1.74 (92); B 20° 1.94 (192); S 1.88$_1$ ±
.015 (204).

C_7H_7ClO o-Chloroanisole, B 10-50° 2.47, Hp -15 to 60° 2.47 (608); B
25° 2.50 (777).
p-Chloroanisole, B 22° 2.24 (155).

$C_7H_7ClO_2S$ Benzylsulfonyl chloride, B 25° 3.85 (848).
p-Toluenesulfonyl chloride, B 20° 5.01 (501); B 5.01 (718, 719);
B 25° 5.00 (848).

$C_7H_7Cl_2I$ o-Tolyliodo dichloride, B 25° 2.55 (643).
m-Tolyliodo dichloride, B 25° 2.82 (643).
p-Tolyliodo dichloride, B 25° 3.02 (643).

C_7H_7F Benzyl fluoride, B 25° 1.77 (857).
C_7H_7FO o-Fluoroanisole, B 25° 2.31 (777).
C_7H_7I o-Iodotoluene, B 22° 1.21 (313).
m-Iodotoluene, B 22° 1.57$_5$ (313).
p-Iodotoluene, B 22° 1.71 (313).

C_7H_7NO Benzamide, D 20° 3.6 (284); D 30″ 3.77 (779).
p-Nitrosotoluene, 3.79 (743).

$C_7H_7NO_2$ o-Aminobenzoic acid, D 1.51 (673).
m-Aminobenzoic acid, D 2.70 (673).
p-Aminobenzoic acid, D 3.51 (673).
o-Nitrotoluene, B 3.56 (17); B 25° 3.75 (64); B 3.64 (79); B 20°
3.69 (192); B 22° 3.66 (313); B 20° 4.22 (421); S 3.66 (657); 30°
B 3.79, T 3.71, Hx 3.91, Hp 3.94, CT 3.81, CD 3.61, Cf 3.01
(737).
m-Nitrotoluene, B 25° 3.81 (64); B 20° 4.17 (192); B 22° 4.14
(313); S 4.17 (657); 30° B 4.29, T 4.12, Hx 4.33, Hp 4.38, CT
4.23, CD 4.01, Cf 3.32 (737).
p-Nitrotoluene, B 4.30 (17); B 25° 4.50 (64); B 4.31 (79); B 20°
4.44 (192); B 22° 4.42 (313); B 25° 4.44 (372); S 4.44 (657); 30°
B 4.52, T 4.41, Hx 4.68, Hp 4.60, CT 4.45, CD 4.29, Cf 3.59
(737).
Phenylnitromethane, B 25° 3.30 (471).

$C_7H_7NO_3$ o-Nitroanisole, B 4.80 (79); B 4.81$_4$ (224); B 20° 4.83 (281);
477°K 4.78 (538).
m-Nitroanisole, B 20° 3.86 (281); 476°K 4.51 (538).
p-Nitroanisole, B 4.36 (79); B 4.75$_3$ (224); B 20° 4.74 (281).

$C_7H_7N_3$ p-Tolyl azide, B 25° 1.96 (189, 316).

$C_7H_7N_3O_2$	Nitroformaldehyde-phenylhydrazone, α-, or cis-, B 20° 3.34 (595).
C_7H_8	Toluene, Lqd. 18° 0.63 (4); CT 25° 0.40 (24); B 25° 0.06 (31); B 0.5 (71); B or Hx 20° 0.39 (192); B 22° < 0.49 (252); E 25° 0.3 (298, 358); 357-482°K 0.37 (308); B 25° 0.34 (428); Hx 20° 0.53 (557); S 20° 0.3 (559); 349-456°K 0.37 (627).
$C_7H_8Br_2O$	1,1-Dimethyl-2,3-dibromo-2-cyclopentene-4-one, B 20° 3.65 (571).
$C_7H_8N_2O$	p-Aminobenzamide, D 20° 4.7 (284).
	Benzoic acid hydrazide (benzoyl hydrazide), B 25° 2.70 (532).
	N-Nitroso-N-methylaniline, B 20° 3.62 (280).
	p-Nitroso-N-methylaniline, B 25° 7.38 (240).
	Phenylurea, D 17° 3.6 + .2 (578).
$C_7H_8N_2O_2$	3-Nitro-4-aminotoluene, B 25° 4.37 (747).
$C_7H_8N_2O_4S$	2-Methyl-5-nitrobenzenesulfonamide, D 25° 4.48 (849).
	p-Nitrotoluene-o-sulfonamide, D 4.51 (718, 719).
C_7H_8O	Anisole, Solid 0.8, B 20-50° 0.8 (40); B 1.16 (79); S 22° 1.23 ± .01 (108); S 1.23 (148); Solid 0 (305); 403°K 1.35 (538); Lqd. 30-40° 1.22 (815).
	Benzyl alcohol, B 1.68 (219); B or Hx 1.69 (237).
	o-Cresol, B 25° 1.44 (61); B 1.41₁ (224); S 1.41 (657); Lqd. 25-30° 2.30 (816).
	m-Cresol, B 25° 1.60 (61); B 1.54₃ (224); S 1.54 (657); Lqd. 25-30° 2.37 (816).
	p-Cresol, B 25° 1.64 (61); B 1.57₃ (224); S 1.57 (657).
C_7H_8OS	2,6-Dimethyl-γ-thiopyrone, B 20° 5.05 (302).
$C_7H_8O_2$	2,6-Dimethyl-γ-pyrone, B 20° 4.05 (302); B 10° 4.47, 20° 4.48, 30° 4.51, 40° 4.51 (506); S 25° 4.6₅ (552); B 15-65° 4.62 ± .02 (573).
	Guaiacol, S 2.46 (620); B or D 25° 2.41 (809).
	o-Hydroxybenzyl alcohol, 25° B 2.64, D 2.78 (809).
C_7H_8S	Thioanisole, B 21° 1.27 (103).
C_7H_9ClO	n-Butylpropiolyl chloride, B 25° 3.06 (545).
C_7H_9N	n-Butylpropiolnitrile, B 25° 4.21 (529).
	2,6-Dimethylpyridine, B 25° 1.65 (416).
	Methylaniline, B 25° 1.64 (163).
	o-Toluidine, B 20° 1.58 (193); B 1.57₆ (224).
	m-Toluidine, B 20° 1.44 (193); B 1.43₂ (224).
	p-Toluidine, B 20° 1.31 (193); B 1.27₃ (224).
C_7H_9NO	o-Anisidine, B 25° 1.50 (145); B 1.45₉ (224); 423-598°K 1.62 (688).
	p-Anisidine, B 25° 1.80 (145); B 1.87₄ (224).
$C_7H_9NO_2S$	Benzylsulfonamide, 25° B 4.02, D 4.63 (849).
	p-Toluenesulfonamide, B 5.02 (718, 719); 25° B 5.00, D 5.39 (849).
$C_7H_{10}N_2$	Heptanedinitrile, B 25° 4.10, 75° 4.26 (569, 669).
	α,α-Phenylmethylhydrazine, B 19° 1.79 (275).
	o-Tolylhydrazine, B 18° 1.54 ± .08 (457).
	m-Tolylhydrazine, B 15° 1.64 ± .06 (457).
	p-Tolylhydrazine, B 18° 2.04 ± .08 (457).
$C_7H_{10}O$	n-Butylpropiolaldehyde, B 25° 3.17 (533).
$C_7H_{10}O_4$	Dimethyl citraconate (cis), B 20° 2.6₈ (580); dimethyl mesaconate (trans), B 20° 2.0₆ (580).
$C_7H_{10}O_7$	Dimethyl citraconate ozonide (cis), B 20° 2.8 (580); dimethyl mesaconate ozonide (trans), B 20° 2.5 (580).
$C_7H_{11}Br$	1-Bromo-1-heptyne, B 25° 1.05 (376).
$C_7H_{11}Cl$	1-Chloro-1-heptyne, B 25° 1.27 (376).

$C_7H_{11}I$ 1-Iodo-1-heptyne, B 25° 0.80 (376).

C_7H_{12} n-Amylacetylene, 348°K 0.878, 398° 0.847 (596).

$C_7H_{12}O_2$ Cyclohexanecarboxylic acid, B 25° 0.9 (143).

$C_7H_{12}O_4$ Diethyl malonate, B 25° 2.54, 50° 2.57 (185); Lqd. 25-30° 2.57 (815).

cyclic-Dioxene trimethylene ether, MP111°, S 0.85 (354); MP158°, S 0.96 (354).

C_7H_{14} Methylcyclohexane, B 25° 0 (143); 370-456°K 0 (627).

$C_7H_{14}Br_2$ 1,2-Dibromo-n-heptane, B 25° 1.76 (386).

2,3-Dibromo-n-heptane, B 25° 2.13 (386).

3,4-Dibromo-n-heptane, B 25° 2.13 (386).

$C_7H_{14}O$ Cyclohexyl methyl ether, 406-473°K 1.29 (538).

Diisopropyl ketone, 25° B 2.740, Hx 2.766, CD 2.663 (540).

Di-n-propyl ketone, B 15° 2.73 (66).

n-Heptaldehyde, B 22° 2.56 (112).

2-Heptanone (methyl amyl ketone), B 22° 2.59 (112).

3-Heptanone (ethyl butyl ketone), B 22° 2.78 (112).

4-Heptanone (dipropyl ketone), B 22° 2.72 (112).

2-Methylcyclohexanol, B 25° 1.95 (143); Lqd. 30-45° 2.47 (816).

3-Methylcyclohexanol, B 25° 1.9 (143); Lqd. 30-35° 2.49 (816); cis-, S 1.75 (662); trans-, S 1.91 (662).

4-Methylcyclohexanol, B 18° 1.71 (117, 234); B 25° 1.9 (143); Lqd. 30-35° 2.68 (816).

$C_7H_{14}O_2$ n-Butyl propionate, B 22° 1.77₈ ± .02 (203).

n-Amyl acetate, B 25° 1.9₁ (127); 376-517°K 1.70 (272).

Isoamyl acetate, B 22° 1.82₃ ± .02 (203); Lqd. 30-40° 1.76 (815).

$C_7H_{15}Br$ 1-Bromoheptane, B 20° 1.8 (72); S 1.86 (89); B 22° 1.85 (112); Hp -90 to 70° 1.81 (138); Solid 0 (305); 373-434°K 2.15 (450).

2-Bromoheptane, B 20° 2.1 (72); B 22° 2.06 (112).

3-Bromoheptane, B 20° 2.0 (72); B 22° 2.04 (112).

4-Bromoheptane, B 20° 2.0 (72); B 22° 2.04 (112).

$C_7H_{15}BrO$ 1-Bromo-2-ethoxypentane, B 25° 2.30 (386).

2-Bromo-3-ethoxypentane, B 25° 2.05 (386).

3-Bromo-2-ethoxypentane, B 25° 2.13 (386).

$C_7H_{15}Cl$ 1-Chloroheptane, B 22° 1.85 (112).

2-Chloroheptane, B 20° 2.0 (72); B 22° 2.03 (112).

3-Chloroheptane, B 20° 2.1 (72); B 22° 2.04 (112).

4-Chloroheptane, B 20° 2.0 (72); B 22° 2.04 (112).

$C_7H_{15}I$ 1-Iodoheptane, B 22° 1.84 (112); CT 20° 1.89 (750).

3-Iodoheptane, B 22° 1.93 (112).

$C_7H_{15}NO_2$ Ethyl α-aminoisovalerate (α-aminovaline ethyl ester), B 25° 2.11 (330).

Ethyl α-aminovalerate, B 25° 2.13 (330).

C_7H_{16} 2,2-Dimethylpentane, Lqd. -120 to 90° 0 (53).

2,3-Dimethylpentane, Lqd. 20 and 30° 0 (53).

2,4-Dimethylpentane, Lqd. 20 and 30° 0 (53).

3,3-Dimethylpentane, Lqd. 20 and 30° 0 (53).

3-Ethylpentane, Lqd. -120 to 90° 0 (53).

n-Heptane, Lqd. -120 to 90° 0 (53); Lqd. -90 to 90° ca. 0 (110); 338-558°K 0 (389).

2-Methylhexane, Lqd. 20 and 30° 0 (53).

3-Methylhexane, Lqd. 20 and 30° 0 (53).

2,2,3-Trimethylbutane, Lqd. 20 and 30° 0 (53).

$C_7H_{16}O$ Ethyl isoamyl ether, B 20-50° 1.15 (40).

1-Heptanol, B 20° 1.7 (72); B 22° 1.71 (112); B 30° 1.66 (430).

2-Heptanol, B 20° 1.7 (72); B 22° 1.71 (112).

3-Heptanol, B 20° 1.7 (72); B 22° 1.71 (112).

4-Heptanol, B 20° 1.7 (72); B 22° 1.70 (112).

$C_8H_4BrNO_2$ p-Nitrophenylbromoacetylene, B 25° 3.42 (398).

C_8H_4ClI o-Chlorophenyliodoacetylene, B 25° 1.42 (398).

p-Chlorophenyliodoacetylene, B 25° 1.06 (398).

$C_8H_4Cl_2N_2$ 2,3-Dichloroquinoxaline, B 2.8 and 3.6 (273); B 3.2 (335).

$C_8H_4Cl_2O_2$ sym-Phthalyl dichloride, B 20° 5.12 (500).

$C_8H_4INO_2$ p-Nitrophenyliodoacetylene, B 25° 3.80 (398).

$C_8H_4N_2$ p-Dicyanobenzene, B 25° 0 (459).

p-Diisocyanobenzene, B 25° 0 (248).

$C_8H_4N_2SSe$ p-Selenocyanophenyl thiocyanate, B 25° 4.02 (863).

$C_8H_4O_3$ Phthalic anhydride, B 10° 5.21, 20° 5.25, 30° 5.27, 40° 5.28 (563).

C_8H_5Br o-Bromophenylacetylene, S 1.79 (373).

m-Bromophenylacetylene, S 1.35 (373).

p-Bromophenylacetylene, S 0.95 (373).

Phenylbromoacetylene, B 25° 0.85 (398).

C_8H_5Cl o-Chlorophenylacetylene, B 25° 1.69 (373, 398).

m-Chlorophenylacetylene, S 1.38 (373).

p-Chlorophenylacetylene, B 25° 0.96 (373, 398).

Phenylchloroacetylene, B 25° 1.10 (398); B 19° 1.21 (461).

$C_8H_5Cl_5$ Ethylpentachlorobenzene, B 25° 0.88 (701).

C_8H_5I Phenyliodoacetylene, B 25° 0.55 (398); B or CT 30° 0.63 (590).

$C_8H_5NO_2$ Isatin, B 20° 5.72 (468).

p-Nitrophenylacetylene, S 13° 3.63 (217); B 25° 3.42 (373, 398).

Phthalimide, B 20° 2.10 (468).

C_8H_6 Phenylacetylene, Hp 10-70° 0.83 (181); S 13° 0.66 (217); B 25° 0.66 (398); B 25° 0.78 (796).

C_8H_6BrN α-Bromobenzyl cyanide, B 20° 3.37 (609).

$C_8H_6Cl_2O_4$ 2,3-Dichloro-5,6-dimethoxybenzoquinone, B 30° 2.99 (779).

3,6-Dichloro-2,5-dimethoxybenzoquinone, B 30° 2.09 (779).

$C_8H_6Cl_4$ Tetrachloro-o-xylene, B 25° 2.65 (701).

$C_8H_6N_2O$ Phenyl furazan, S 3.997 (433).

$C_8H_6N_2O_2$ p-Nitrobenzyl cyanide, B 25° 3.84, 50° 3.95 (258).

C_8H_6O Benzofuran (coumarone), S 0.79 (620).

Phenoxyacetylene, B 25° 1.41 (796).

$C_8H_6O_2$ o-Phthalic aldehyde (phthalaldehyde), B 4.50 (119).

m-Phthalic aldehyde (isophthalaldehyde), B 2.86 (119).

p-Phthalic aldehyde (terephthalaldehyde), B 2.35 (119).

C_8H_7Br p-Bromophenylethylene (4-bromostyrene), B 25° 1.35 (439).

1-Phenyl-2-bromoethylene (β- or ω-bromostyrene), B 18° 1.51 (461); high melting, 30° B 1.54, CT 1.53 (590); low melting, 30° B 1.12, CT 1.38 (590).

C_8H_7BrO ω-Bromoacetophenone (bromomethyl phenyl ketone), B 20° 3.11 (609).

p-Bromoacetophenone, S 2.29 (233).

$C_8H_7BrO_2$ Methyl p-bromobenzoate, B 14° 1.82 (461).

C_8H_7Cl β-Chlorostyrene (ω-chlorostyrene), B 19° 1.40 (461).

4-Chlorostyrene (p-chlorophenylethylene), B 25° 1.28 (439).

C_8H_7ClO ω-Chloroacetophenone, B 20° 3.26 (609).

p-Chloroacetophenone, S 2.29 (233); S 1.57 ± .03 (293).

Phenylacetyl chloride, B 20° 2.54 (499).

p-Toluoyl chloride, B 20° 3.81 (499).

$C_8H_7Cl_3$ 3,4,5-Trichloro-o-xylene, B 25° 2.46 (701).

C_8H_7IO p-Iodoacetophenone, S 2.23 (233).

C_8H_7N Benzyl cyanide, B 25° 3.47, 50° 3.48, Hp 25° 3.55, 50° 3.56 (258).

Indole, B 20° 2,05 (468); B 25° 2,11 (797).

o-Methoxybenzonitrile, B 25° 4,97 (847).

o-Toluisonitrile, B 22° 3,35 (252).

p-Toluisonitrile (p-tolyl isocyanide), B 25° 3,98 ± .03 (115);
B 22° 3,95₅ (252).

o-Tolunitrile, S 3,77₅ ± .01 (204).

m-Tolunitrile, B 22° 4,18 (252).

p-Tolunitrile, B 22° 4,37 (252).

C_8H_7NOSe	p-Methoxyphenyl selenocyanate, B 25° 4,42 (863).
$C_8H_7NO_2$	β- or ω-Nitrostyrene, B 17° 4,48 (461); D 25° 4,27 (589).
C_8H_7NS	p-Methylphenyl isothiocyanate, B 21° 3,32 (215).
$C_8H_7NS_2$	2-Methylmercaptobenzothiazole, B 25° 1,42 (768).

4-Methyl-2-mercaptobenzothiazole, B 25° 4,00 (768).

6-Methyl-2-mercaptobenzothiazole, B 25° 4,30 (768).

N-Methylbenzothiazolethione, B 25° 4,84 (768).

C_8H_7NSe	Benzyl selenocyanate, B 25° 3,98 (860).
	p-Tolyl selenocyanate, B 25° 4,35 (860).
$C_8H_7N_3$	1-Phenyl-1,2,3-triazole, B 25° 4,08 (787).

1-Phenyl-1,2,4-triazole, B 25° 2,88 (787).

2-Phenyl-1,2,3-triazole, B 25° 0,97 (787).

4-Phenyl-1,2,4-triazole, B 25° 5,63 (787).

C_8H_8	Styrene (phenylethylene), B 25° 0,56 (164); B 10-70° 0 (181);

B 25° 0,37 (439); S 20° 0,3 (559); B or CT 30° 0,37 (590); S
0,31 (685); 442-462°K < 0,2 (829).

$C_8H_8Br_2$	4,5-Dibromo-o-xylene, B 25° 2,86 (507).

o-Xylylene dibromide, B 25° 2,02, 50° 2,08, Hx 20° 1,98, 40°
2,04, CT -15° 1,79, 0° 1,83, 15° 1,87, 30° 1,88, 45° 1,92 (546).

p-Xylylene dibromide, B 22° 2,04 (640).

$C_8H_8Cl_2$	4,5-Dichloro-o-xylene, B 25° 3,01 (701).
	p-Xylylene dichloride, B 25° 2,2₃ (94); B 22° 2,17 (640).
$C_8H_8N_2O$	1,3-Dimethyl-o-benzoquinone furazan, D 25° 4,82 (741).
$C_8H_8N_2O_2$	1,3-Dimethyl-o-benzoquinone furoxan, (1,3-dimethyl-o-quinone

dioxime peroxide), D 25° 5,40 (694, 741).

Phenylglyoxime, α -, D 20° 1,396 (366); β-, D 20° 1,702 (366).

$C_8H_8N_2O_3$	p-Nitrobenzaldoxime-N-methyl ether, B 25° 6,4 ± .1 (191).

p-Nitrobenzaldoxime-O-methyl ether, α-, S 25° 3,39₅ (323);
β-, S 25° 3,88₅ (323).

C_8H_8O	Acetophenone, B 18° 2,97 (77); 19° B 2,97, T 2,87, Cf 2,81

(116, 121); B 2,97, T 2,87, Cf 2,75 (122); S 2,85 (233); B 2,90 ±
.02 (293); 410-493°K 3,00 (417); B 25° 2,77 (533); Lqd. 30-40°
3,67 (815).

Phenylacetaldehyde, B 20° 2,48 (583).

Styrene oxide, B 12° 1,64 (213).

p-Tolualdehyde, S 25° 3,26 (311); D 25° 3,27 (375); B 25° 3,30
(443).

$C_8H_8O_2$	2,5-Dimethyl-1,4-benzoquinone, B 25° 0,68 (418).

o-Hydroxyacetophenone, 25° B 3,16, D 3,23 (809).

Methyl benzoate, B 20-50° 1,8 (40); B 1,8 (42); B 22° 1,83₄ ±
.02 (161); CT 1,9 (208); B 25° 2,52 (454); Lqd. 30-40° 1,83
(815).

o-Methoxybenzaldehyde, B 25° 4,21 (809).

p-Methoxybenzaldehyde (p-anisaldehyde), S 25° 3,70 (311); D
25° 3,70 (375); B 25° 3,85 (443).

Phenyl acetate, B 22° 1,52₇ ± .015 (161); CT 1,4 (208).

Phenylacetic acid, D 25° 1,75 (460).

cyclic-Phenylene ethylene ether, S 1,40 (354).

$C_8H_8O_3$	Methyl salicylate, B 13 and 40° 2,41 (301, 304); Lqd. 30-40°

	$\underline{2.34}$ (816).
	Vanillin, S $\underline{3.0}$ (620).
$C_8H_8O_4$	Dehydroacetic acid, S 25° $\underline{2.8}_3$ (552).
C_8H_9Br	4-Bromo-o-xylene, B 25° $\underline{2.07}$ (507).
	ω-Bromo-p-xylene, Hx 20° $\underline{2.07}$ (609).
C_8H_9BrO	p-Bromophenetole, B 25° $\underline{2.38}$, 50° $\underline{2.41}$, Hp 25° $\underline{2.46}$, 50° $\underline{2.48}$ (260).
C_8H_9ClO	o-Chlorophenetole, B 25° $\underline{2.54}$ (777).
C_8H_9NO	Acetanilide, B 25° $\underline{4.01}$ (492).
	m-Aminoacetophenone, B 18° $\underline{5.4}$ (708).
	p-Aminoacetophenone, S $\underline{4.29}$ (233); S $\underline{4.3}$ (772).
	O-Methylbenzaldoxime, anti-, B 18° $\underline{0.86}$ (77).
$C_8H_9NO_2$	Ethyl isonicotinate, B 25° $\underline{2.49}$ (814).
	Methyl o-aminobenzoate, B 20 and 50° $\underline{1.0}$ (40); ca. 373°K $\underline{1.0}$ (41).
	Methyl m-aminobenzoate, B 20-50° $\underline{2.4}$ (40); ca. 373°K $\underline{2.4}$ (41).
	Methyl p-aminobenzoate, B 20 and 50° $\underline{3.3}$ (40); ca. 373°K $\underline{3.3}$ (41).
$C_8H_9NO_4$	Nitrohydroquinone dimethyl ether, B 18° $\underline{4.56}$ (78).
C_8H_{10}	Ethylbenzene, 349-455°K $\underline{0.58}$ (627); Lqd. 20° $\underline{0.35}$ (749).
	o-Xylene, B 25° $\underline{0.52}$ (31); Lqd. 15-45° $\underline{0.51}$ (177); B 20° $\underline{0.58}$ (192); Lqd. -20 to 130° $\underline{0.44}$ (235); S 20° $\underline{0.5}$ (559); S $\underline{0.55}$ (657); Vap. $\underline{0.62}$ (757).
	m-Xylene, Lqd. 15-45° $\underline{0.36}$ (177); B 20° $\underline{0.37}$ (192); Lqd. -40 to 120° $\underline{0.34}$ (235); S $\underline{0.4}$ (657).
	p-Xylene, B 25° $\underline{0.06}$ (31); B < $\underline{0.1}$ (60); Lqd. 15-45° $\underline{0.17}$ (177); B 20° $\underline{0.1}_2$ (192); Hp 20° $\underline{0}$ (220); S $\underline{0}$ (657); Lqd. 20° $\underline{0}$ (749); Vap. $\underline{0}$ (757).
$C_8H_{10}BrClO_2$	2,2-Chlorobromomethone, CT 19° $\underline{4.36}$ (571, 622).
$C_8H_{10}BrN$	p-Bromo-N-dimethylaniline, B 25° $\underline{3.42}$ (420); B 25° $\underline{3.37}$ (495).
$C_8H_{10}Br_2O_2$	2,2-Dibromomethone, eth. alc. 18° $\underline{4.29}$ (571); CT $\underline{4.29}$ (622).
$C_8H_{10}ClN$	p-Chloro-N-dimethylaniline, B 25° $\underline{3.29}_5$ (495).
$C_8H_{10}Cl_2O_2$	2,2-Dichloromethone, B 21° $\underline{4.49}$ (571, 622).
$C_8H_{10}IN$	p-Iodo-N-dimethylaniline, B 25° $\underline{3.24}$ ± .02 (495).
$C_8H_{10}N_2$	Acetaldehydephenylhydrazone, a- or cis-, B 20° $\underline{2.61}$ (595); β- or anti-, B 20° $\underline{2.58}$ (595).
$C_8H_{10}N_2O$	p-Nitroso-N-dimethylaniline, B 25° $\underline{6.89}$, 45° $\underline{6.91}$, CT 25° $\underline{6.33}$ (240); B 25° $\underline{6.75}$ (420); B $\underline{6.76}$ (678).
	N-Nitroso-N-ethylaniline, B 20° $\underline{3.61}$ (280).
$C_8H_{10}N_2O_2$	p-Nitro-N-dimethylaniline, B 25° $\underline{6.15}$ (420); B 25° $\underline{6.87}$ (495).
	2-Nitro-5-m-xylidine (2-nitro-5-amino-1,3-dimethylbenzene), B 25° $\underline{5.04}$ (648).
$C_8H_{10}O$	o-Cresyl methyl ether (o-tolyl methyl ether), B 25° $\underline{1.0}$ (61).
	m-Cresyl methyl ether, B 25° $\underline{1.17}$ (61).
	p-Cresyl methyl ether, B 25° $\underline{1.20}$ (61).
	Phenethyl alcohol (2-phenylethanol), B 25° $\underline{1.64}$ (80).
	Phenetole, Lqd. 20-50° $\underline{0.7}$, B 20-50° $\underline{1.0}$ (40); 415 and 473°K $\underline{1.40}$ (538); Lqd. 30-40° $\underline{1.27}$ (815).
	Phenylmethylcarbinol, B $\underline{1.60}$, Hx $\underline{1.55}$ (219).
$C_8H_{10}O_2$	o-Dimethoxybenzene (veratrole), S 10° $\underline{1.18}$, 25° $\underline{1.24}$, 40° $\underline{1.32}$ (608); D 25° $\underline{1.23}$ (809).
	m-Dimethoxybenzene (resorcinol dimethyl ether), B 25° $\underline{1.5}_8$ (94); S 10° $\underline{1.57}$, 25° $\underline{1.58}$, 40° $\underline{1.5}_9$ (608).
	p-Dimethoxybenzene (hydroquinone dimethyl ether), B 18° $\underline{1.74}$ (78); B 25° $\underline{1.81}$ (94); B 18-40° $\underline{1.67}$ (116, 120); Solid $\underline{0}$ (305); S 10° $\underline{1.69}$, 25° $\underline{1.70}$, 40° $\underline{1.71}$ (608).
$C_8H_{10}O_3S$	Methyl p-toluenesulfonate (methyl ester of tosylic acid), B 25°

5.18 (602); B 20° 4.85 (676).

$C_8H_{11}BrO_2$ 2-Bromomethone, eth. alc. 22° 3.81 (571, 622).

$C_8H_{11}ClO$ n-Amylpropiolyl chloride, B 25° 3.08 (545).

$C_8H_{11}N$ n-Amylpropiolnitrile, B 25° 4.22 (529).

N-Dimethylaniline, B 1.39 (17); B 25° 1.58 (163); B 25° 1.55 (420); B 25° 1.58 (495); 455°K 1.61 (538).

N-Ethylaniline, B 20° 1.68 (280).

2,4,6-Trimethylpyridine, B 10-40° 1.93 ± .01 (382).

$C_8H_{11}NO$ N-Dimethylaniline oxide, 25° B 4.79, D 4.85 (689).

$C_8H_{11}NO_2S$ N-Methyl-p-toluenesulfonamide, B 25° 5.4 (602).

$C_8H_{12}N_2$ p-Amino-N-dimethylaniline, B 25° 1.42 (495).

Tetramethylpyrazine, B 0 (273); B < 0.45 (335).

$C_8H_{12}O$ n-Amylpropiolaldehyde, B 25° 3.18 (533).

n-Butylacetylacetylene, B 25° 3.20 (533).

3,5-Dimethyl-2-cyclohexene-1-one, B 18° 3.79 (234).

$C_8H_{12}O_2$ Ethyl sorbate, 507°K 2.07 (831).

Methone (5,5-dimethyl-1,3-cyclohexanedione), eth. alc. 21° 3.57 (571, 622).

Tetramethylcyclobutane-1,3-dione, CT 25° ca. 0 (401); B 25° 0.72 (418).

$C_8H_{12}O_4$ Diethyl maleate (cis), B 25° 2.54, 50° 2.56 (185); B 25 and 50° 2.54 (303); CT 23° 2.51 (524); S 2.55 (657); diethyl fumarate (trans), B 25° 2.38, 50° 2.40 (185); B 25 and 50° 2.38 (303); CT 23° 2.23 (524); S 2.39 (657).

$C_8H_{12}O_7$ Diethyl maleate ozonide (cis), CT 23° 2.53 (524); diethyl fumarate ozonide (trans), CT 23° 2.29 (524).

$C_8H_{14}O$ 3,3-Dimethylcyclohexanone, B 18° 2.92 (234).

3,4-Dimethylcyclohexanone, B 18° 2.83 (234).

3,5-Dimethylcyclohexanone, B 18° 2.89 (234).

$C_8H_{14}O_2$ Cyclohexyl acetate, B 18° 1.90 (234).

$C_8H_{14}O_4$ 2,3-Diacetoxybutane, dl-, Lqd. extrapolated to 25° 1.95 (710); meso-, Lqd. 25° 2.35 (710).

Diethyl succinate, S 2.19 (117); kerosene 0° 2.01, 20° 2.08, 40° 2.14, 60° 2.19, 80° 2.25, 100° 2.30, 120° 2.34, 140° 2.39, 160° 2.43, 180° 2.47 (182, 183); B 25° 2.14, 50° 2.21 (185); 430-519°K 2.3$_0$ (270); Lqd. 30-40° 2.37 (815).

$C_8H_{14}O_6$ Diethyl tartrate, dl-, B 12° 3.09 ± .02, 22° 3.12 ± .025, 38° 3.16 ± .03 (148); B 25° 3.10 (202); meso-, B 12 and 22° 3.66 ± .04, 38° 3.69 ± .04 (148); B 25° 3.66 (202).

C_8H_{16} Ethylcyclohexane, 370-456°K 0 (627).

$C_8H_{16}O$ 3,5-Dimethylcyclohexanol, 3-cis, 5-cis-, 1-cis-, S 20° 1.31 (662); 3-cis-, 5-cis-, 1-trans-, S 20° 1.82 (662); 3-cis-, 5-trans-, 1-cis-, S 20° 1.85 (662).

α-Ethylcaproaldehyde, B 25° 2.64 (443).

Methyl hexyl ketone, B 15° 2.70 (546).

$C_8H_{16}O_2$ 2,4,4,5,5-Pentamethyl-1,3-dioxolane, B 25° 1.29 (560).

$C_8H_{17}Br$ n-Octyl bromide, B 14° 1.96 (646).

$C_8H_{17}I$ 1-Iodooctane, CT 20° 1.89 (750).

2-Iodooctane, CT 20° 2.06 (750).

$C_8H_{17}NO$ 2-Nitroso-2,5-dimethylhexane, B 25° 2.51 (231); CT 0° 0.99 ± .10 (351).

$C_8H_{17}NO_2$ Ethyl α-aminoisocaproate (leucine ethyl ester), B 25° 2.03 (330).

C_8H_{18} n-Octane, Lqd. -50 to 110° ca. 0 (110).

2,2,4-Trimethylpentane, Lqd. -100 to -10° 0 (53).

$C_8H_{18}O$ n-Butyl ether, 385-455°K 1.18 (537); 25° B 1.22, T 1.13, kerosene 1.17 (836).

2-Ethyl-1-hexanol, B 25° 1.74 (443).

2-Methyl-3-heptanol, B 20-70° 1.62 (91).

n-Octyl alcohol, B 25° 1.62 (74); B 25° 1.64 (82); B 20-70° 1.70 (90); Lqd. 20° 3.02 (816).

$C_8H_{18}OS$ Isobutyl sulfoxide, B 25° 3.90 (591).

$C_8H_{18}S$ n-Butyl sulfide, B 20° 1.57 (238); B 20° 1.56 (422).

$C_8H_{20}N_2$ Octamethylenediamine, B 25-75° 1.99 (669).

$C_8H_{20}N_2OS$ Sulfurous acid bis-diethylamide, B 25° 3.00 (728).

$C_9H_5ClO_2$ o-Chlorophenylpropiolic acid, D 25° 2.63 (460).

p-Chlorophenylpropiolic acid, D 25° 1.95 (460).

C_9H_5N Phenylpropiolnitrile, B 25° 4.50 (529).

$C_9H_5NO_4$ o-Nitrophenylpropiolic acid, D 25° 4.02 (460).

C_9H_6O 2,3-Diphenylindone, B 3.28 (104).

Phenylpropiolaldehyde, B 25° 3.36 (533).

$C_9H_6O_2$ Coumarin, B 10-40° 4.50 to 4.52 (505, 506); S 25° 4.4$_8$ (552); B 15-65° 4.54 ± .02 (573).

Phenylpropiolic acid, D 25° 2.29 (460).

C_9H_7Br p-Tolylbromoacetylene, B 25° 1.27 (398).

C_9H_7Cl p-Tolylchloroacetylene, B 25° 1.90 (398).

C_9H_7ClO Cinnamoyl chloride, B 25° 3.63 (545).

C_9H_7I p-Tolyliodoacetylene, B 25° 0.97 (398).

C_9H_7N Cinnamic acid nitrile (β-cyanostyrene), D 25° 4.14 (589).

Isoquinoline, B 14° 2.53 (211); B 25° 2.524 (241); B 10-40° 2.54 (382).

Quinoline, B 25° 2.25 (47); B 14° 2.14 (211); B 25° 2.184 (241); B 10-40° 2.16 (382).

C_9H_7NO 8-Hydroxyquinoline, B or Hp 2.70 (237).

α-Phenylisoxazole, B 3.34 (704).

C_9H_8 Indene, S 0.44 (685); B 0.67 (806).

p-Tolylacetylene (p-methylphenylacetylene), B 25° 1.01 (373, 398).

$C_9H_8Br_2$ 5,6-Dibromohydrindene, B 25° 2.48 (507).

$C_9H_8Br_2O_2$ 2,2-Dimethyl-5,6-dibromo-o-phenylene methylene dioxide, S 3.28 (466).

$C_9H_8Cl_2O_2$ 2,2-Dimethyl-5,6-dichloro-o-phenylene methylene dioxide, S 3.39 (466).

$C_9H_8N_2$ 2- or α-Methylquinazoline, B 1.6 and 2.4 (273); B 25° 2.2 (390).

2-Methylquinoxaline, B 2.2 (335).

$C_9H_8N_2O$ 3-Methyl-5-phenylazoxime, S 1.63 (433).

5-Methyl-3-phenylazoxime, S 1.41 (433).

Methylphenylfurazan, S 4.141 (433).

Methylphenyloxydiazole, S 3.434 (433).

p-Tolylfurazan, S 4.176 (433).

$C_9H_8N_2O_2$ Methylphenylfurazan peroxide, B 25° 4.765 (666).

Methylphenylfuroxan, B 25° 4.92 (666).

$C_9H_8N_2S_2$ 3-Phenyl-2-methyl-2,5-endothio-1,3,4-thiadiazoline, D 25° 8.8 (787).

C_9H_8O Cinnamic aldehyde, B 18° 3.71 (120).

1- or α-Indanone (1- or α-hydrindone), 20° B or CH 3.3, CT 3.5 (795).

$C_9H_8O_2$ Cinnamic acid, D 25° 1.78 (589).

C_9H_9Br 6-Bromohydrindene, B 25° 2.15 (507).

o-Bromo-α-methylstyrene, B 27° 1.87 (464).

p-Bromo-α-methylstyrene, B 22° 1.45 (464).

$C_9H_9BrO_2$ 2,2-Dimethyl-6-bromo-o-phenylene methylene dioxide, S 2.70 (466).

Ethyl p-bromobenzoate, S 30° 2.31 (738).

$C_9H_9Br_3$	Tribromomesitylene, CT 20 and 54° $\underline{0}$ (172); B 40° $\underline{0.35}$ (192).
C_9H_9Cl	m-Chloro-α-methylstyrene, B 27° $\underline{1.89}$ (464).
	Cinnamyl chloride, D 25° $\underline{1.90}$ (589).
$C_9H_9ClO_2$	2,2-Dimethyl-6-chloro-o-phenylene methylene dioxide, S $\underline{2.74}$ (466).
	Ethyl p-chlorobenzoate, B 20° $\underline{2.00}$ (155); S 30° $\underline{2.24}$ (738).
$C_9H_9Cl_3$	Trichloromesitylene, B 40° $\underline{0.38}$ (192); B 25° ca. $\underline{0.1}$ (701).
	Trichloropseudocumene, B 25° $\underline{1.83}$ (701).
C_9H_9F	o-Fluoro-α-methylstyrene, B 26° $\underline{1.54}$ (464).
C_9H_9I	o-Iodo-α-methylstyrene, B 26° $\underline{1.48}$ (464).
C_9H_9N	1-Methylindole, B 25° $\underline{2.16}$ (797).
	2-Methylindole, B 25° $\underline{2.47}$ (797).
	3-Methylindole (skatole), B 20° $\underline{2.08}$ (468).
C_9H_9NO	Cinnamoylamide, D 25° $\underline{3.61}$ (589).
$C_9H_9NO_2$	p-Methyl-β-nitrostyrene, D 25° $\underline{4.77}$ (589).
$C_9H_9NO_4$	Ethyl p-nitrobenzoate, B 13° $\underline{3.84}$ (461); S 30° $\underline{4.05}$ (738).
$C_9H_9N_3O_2S_2$	2-Sulfanilamidothiazole (sulfathiazole), D 25° $\underline{7.0}$ ± .2 (759).
$C_9H_9N_3O_6$	Trinitromesitylene, B 20 and 51° $\underline{0}$ (172); B 20° $\underline{0.79}$ (192); B 25 and 45° ca. 0.5 (426).
C_9H_{10}	Allylbenzene, B 20° $\underline{0.5}$ (632).
	Hydrindene, B 25° $\underline{0.53}$ (507).
	Isopropenylbenzene (unsym-methylphenylethylene), B 20° ca. $\underline{0.75}$ (632).
	p-Methylphenylethylene, B 25° $\underline{0.63}$ (439).
	Phenylcyclopropane, B 25° $\underline{0.49}$ (859).
	Propenylbenzene (isoallylbenzene), B 20° $\underline{0.71}$ (632).
$C_9H_{10}N_2$	p-Dimethylaminobenzonitrile, B 25° $\underline{5.90}$ (420).
$C_9H_{10}N_2O_2$	Methylphenylglyoxime, D 20° $\underline{1.161}$ (366).
	p-Tolylglyoxime, D 20° $\underline{1.66}$ (366).
$C_9H_{10}N_2S$	p-Dimethylaminophenyl thiocyanate, B 25° $\underline{5.70}$ (863).
$C_9H_{10}N_2Se$	p-Dimethylaminophenyl selenocyanate, B 25° $\underline{5.64}$ (863).
$C_9H_{10}N_4O_2S_2$	2-Sulfanilamido-5-methyl-1,3,4-thiadiazole, D 25° $\underline{6.8}$ ± .2 (759).
$C_9H_{10}O$	Cinnamyl alcohol, D 25° $\underline{1.79}$ (589).
	α-Phenylpropionaldehyde, B 20° $\underline{2.79}$ (583).
	β-Phenylpropionaldehyde, B 20° $\underline{2.31}$, Hx 20° $\underline{2.34}$ (583).
$C_9H_{10}O_2$	Benzyl acetate, B 25° $\underline{1.8}$ (127).
	o-Cresyl acetate, B 22° $\underline{1.67}_8$ ± .022 (161).
	m-Cresyl acetate, B 22° $\underline{1.59}_8$ ± .015 (161).
	p-Cresyl acetate, B 22° $\underline{1.52}_6$ ± .015 (161).
	2,2-Dimethyl-o-phenylene methylene dioxide, S $\underline{1.02}$ (466).
	Ethyl benzoate, B $\underline{1.8}$ (42); B 22° $\underline{1.82}_9$ ± .02 (161); B 25° $\underline{1.91}$ (405); 405-505°K $\underline{1.95}$ (417); B 25° $\underline{2.43}$ (454); CT 20° $\underline{1.99}$ (534); S 30° $\underline{1.93}$ (738); B 30° $\underline{1.88}$ (779); Lqd. 30-40° $\underline{1.87}$ (815).
	Methyl o-toluate, B 22° $\underline{1.59}_8$ ± .015 (161).
	Methyl m-toluate, B 22° $\underline{1.92}_2$ ± .015 (161).
	Methyl p-toluate, B 22° $\underline{2.04}_9$ ± .023 (161).
	cyclic-Phenylene trimethylene ether, S $\underline{1.93}$ (354).
	Phenyl propionate, B 22° $\underline{1.52}_7$ ± .016 (161).
$C_9H_{10}O_3$	Allylbenzene ozonide, B 20° $\underline{1.18}$ (624).
	Ethyl salicylate, B 40° $\underline{2.88}$ (301, 304); Lqd. 30-40° $\underline{2.83}$ (816).
	Methyl mandelate, dl-, B 25° $\underline{2.4}_5$ (200); d-, B 25° $\underline{2.4}_7$ (200).
$C_9H_{11}Br$	Bromomesitylene, B 30° $\underline{1.52}$ (343).
	3-Phenylpropyl bromide, D 25° $\underline{1.78}$ (589).
$C_9H_{11}Cl$	Chloromesitylene, B 30° $\underline{1.55}$ (343).
	3-Phenylpropyl chloride, D 25° $\underline{1.76}$ (589).

$C_9H_{11}F$ **Fluoromesitylene**, B 30° 1.36 (343).
$C_9H_{11}I$ **Iodomesitylene**, B 30° 1.42 (343).
$C_9H_{11}NO$ **p-Dimethylaminobenzaldehyde**, B 18° 5.6 (708).
    Nitrosomesitylene, B 8° 1.37 ± .13, 25° 1.63 ± .09 (351).
$C_9H_{11}NO_2$ **p-Aminobenzyl acetate**, S 2.8 (772).
    N-Dimethylanthranilic acid, B 25° 6.31 (404).
    Ethyl p-aminobenzoate, B 19° 4.0 (708); S 30° 3.41 (738); S 3.14 (772).
    Nitromesitylene, 30° B 3.65, CT 3.63 (343); B 25° 3.67 ± .01 (351).

C_9H_{12} p-Ethyltoluene, B 25° 0 (428).
    Isopropylbenzene (cumene), S 20° 0.95 (559); 369-455°K 0.65 (627).
    Mesitylene, B 25° 0.1 (58); B 25°< 0.2 (61); B 20° 0.0₇ (192).
$C_9H_{12}N_2$ Acetonephenylhydrazone, B 20° 2.68 (595).
$C_9H_{12}O$ 3- or γ-Phenylpropyl alcohol, B 25° 1.63 (80); D 25° 1.71 (589).
    2,4,6-Trimethylphenol (mesitol), B 30° 1.36 (343).
$C_9H_{12}O_3$ Phloroglucinol trimethyl ether, B 25° 1.8 (61).
$C_9H_{12}O_8$ Tetracarboxymethane tetramethyl ester, B > 0 (35).
$C_9H_{13}BrO_2$ 2-Bromo-2-methylmethone, B 21° 2.89 (571, 622).
$C_9H_{13}ClO_2$ 2-Chloro-2-methylmethone, CT 23° 3.00 (571, 622).
$C_9H_{13}N$ N-Dimethyl-p-toluidine, B 25° 1.29 (495).
    2,4,6-Trimethylaniline (mesidine), B 25° 1.40 (648).
$C_9H_{13}NO$ p-Aminophenyl propyl ether, S 1.7 (772).
    N-Dimethyl-p-anisidine, B 25° 1.70 (420).
$C_9H_{13}NO_2S$ N-Dimethyl-p-toluenesulfonamide, B 25° 5.48 (602).
$C_9H_{14}N_2$ Nonanedinitrile, B 25° 4.39, 75° 4.47 (569, 669).
$C_9H_{14}O$ n-Amyl-acetylacetylene, B 25° 3.20 (533).
    Isophorone, D 25° 3.96 (812).
$C_9H_{14}O_4$ Diethyl cyclopropane-1,1-dicarboxylate, B 25° 2.40 (290); B 2.4 (773).
$C_9H_{16}O_4$ 2,3-Diacetoxy-n-pentane, dl-threo-, Lqd. 25° 2.07 (710); dl-erythro-, Lqd. 25° 2.48 (710).
    Diethyl dimethylmalonate, B 2.32 (773).
    Diethyl glutarate, B 25° 2.41, 50° 2.42 (185); Lqd, 30-40° 2.46 (815).
$C_9H_{17}NO_4$ Diethyl glutamate, B 25° 2.56 (330).
C_9H_{18} Isopropylcyclohexane, 391-456°K 0 (627).
$C_9H_{18}Br_2$ Nonamethylene bromide, Hp 25° 2.55, 50° 2.58 (320).
$C_9H_{18}O$ Hexamethylacetone, B 15° 2.76 (66).
    1,3,5-Trimethylcyclohexanol, B 18° 1.86 (234).
$C_9H_{19}Br$ n-Nonyl bromide, B 1.89 (721).
$C_9H_{19}BrO$ 1-Bromo-2-ethoxy-n-heptane, B 20° 2.27 (386).
    2-Bromo-3-ethoxy-n-heptane, B 20° 2.08 (386).
    3-Bromo-4-ethoxy-n-heptane, B 25° 2.11 (386).
$C_9H_{20}N_2O$ Tetraethylurea, B 20° 3.3 (284).
$C_9H_{20}O$ n-Nonyl alcohol, B 25° 1.60 (74, 82).
$C_9H_{20}O_2$ 2,2-Dimethoxyheptane, B 25° 0.90 (437).
$C_9H_{20}O_4$ Pentaerythritol tetramethyl ether, B 0.8 (23).
    Tetraethyl orthocarbonate, B > 0 (35).
$C_{10}H_4Cl_6O_4$ Hydroquinone bis-trichloroacetate, B 25° 2.2 (755).
$C_{10}H_6BrCl$ 1-Chloro-8-bromonaphthalene, B 14° 2.64 (462).
$C_{10}H_6BrF$ 1-Bromo-2-fluoronaphthalene, B 20° 2.34 (174, 249, 435).
$C_{10}H_6BrI$ 1-Bromo-2-iodonaphthalene, B 19° 1.80 (174, 249, 435).
$C_{10}H_6BrNO_2$ 1-Bromo-5-nitronaphthalene, B 22° 2.49 (174, 435); S 2.495 (249).
$C_{10}H_6ClF$ 1-Chloro-8-fluoronaphthalene, B 19° 2.86 (462).

$C_{10}H_6ClI$ 1-Chloro-8-iodonaphthalene, B 14° $\underline{2.55}$ (462).

$C_{10}H_6Cl_2$ 1,2-Dichloronaphthalene, B 25° $\underline{2.4}_4$ (395); B 25° $\underline{2.47}$ (474).

1,3-Dichloronaphthalene, B 25° $\underline{1.8}_5$ (395); B 25° $\underline{1.78}$ (474).

1,4-Dichloronaphthalene, B 19° $\underline{0}$ (174, 249, 435); B 25° $\underline{0.5}_0$ (395).

1,5-Dichloronaphthalene, B 25° $\underline{0}$ (395); B 25° $\underline{0}$ (474).

1,6-Dichloronaphthalene, B 25° $\underline{1.44}$ (474).

1,7-Dichloronaphthalene, B 20° $\underline{2.58}$ (174, 435); B 25° $\underline{2.55}$ (474).

1,8-Dichloronaphthalene, B 25° $\underline{2.8}_8$ (395); B 15° $\underline{2.78}$ (462); B 25° $\underline{2.82}$ (474).

2,3-Dichloronaphthalene, B 25° $\underline{2.55}$ (474).

2,6-Dichloronaphthalene, B 14° $\underline{0.60}$ (174, 249, 435); B 25° $\underline{0}$ (201, 474); B 25° $\underline{0.2}_0$ (327).

2,7-Dichloronaphthalene, B 25° $\underline{1.53}$ (395, 474).

$C_{10}H_6Cl_2O_4S_2$ 1,5-Naphthalenedisulfonyl chloride (1,5-di(sulfochloro) naphthalene), B $\underline{1.66}$ (718, 719); B 25° $\underline{1.55}$ (834).

$C_{10}H_6F_2$ 1,5-Difluoronaphthalene, B 22° ca. $\underline{0}$ (174, 435); $\underline{0}$ (249).

$C_{10}H_6N_2O$ 1,2- or β-Naphthoquinone dioxime furazan, S $\underline{4.84}$ (607); B 25° $\underline{4.3}$ (821).

$C_{10}H_6N_2O_2$ 1,2- or β-Naphthoquinone dioxime peroxide, S $\underline{5.33}$ (607).

$C_{10}H_6N_2O_4$ 1,5-Dinitronaphthalene, B $\underline{0.6}$ (79); S $\underline{0.6}$ (249).

1,8-Dinitronaphthalene, B $\underline{7.1}$ (79); S $\underline{7.1}$ (249); B 14° $\underline{7.87}$ (462); B 25° $\underline{7.22}$ (747).

$C_{10}H_7Br$ 1-Bromonaphthalene, B 20° $\underline{1.58}$ (131); B 20° $\underline{1.48}$ (174, 249, 435).

2-Bromonaphthalene, B 20° $\underline{1.71}$ (131); B 19° $\underline{1.69}$ (174, 249, 435).

$C_{10}H_7BrN_2O_2$ 4-Bromo-2-nitro-1-naphthylamine, D 25° $\underline{5.60}$ (747).

5-Bromo-2-nitro-1-naphthylamine, D 25° $\underline{5.05}$ (747).

6-Bromo-1-nitro-2-naphthylamine, D 25° $\underline{5.13}$ (747).

$C_{10}H_7BrO_2S$ α-Naphthalenesulfonyl chloride, B 25° $\underline{4.92}$ (848).

$C_{10}H_7Cl$ 1-Chloronaphthalene, B 20° $\underline{1.59}$ (131); B 19° $\underline{1.50}$ (174, 249, 435); B 25° $\underline{1.51}$ (474).

2-Chloronaphthalene, B 20° $\underline{1.72}$ (131); B 20° $\underline{1.57}$ (174, 249, 435); B 25° $\underline{1.65}$ (474).

$C_{10}H_7ClO_2S$ 1- or α-Naphthalenesulfonyl chloride, B 25° $\underline{4.76}$ (848).

2- or β-Naphthalenesulfonyl chloride, B $\underline{4.98}$ (718, 719); B 25° $\underline{4.96}$ (848).

$C_{10}H_7F$ 1-Fluoronaphthalene, B 20° $\underline{1.41}$ (131); B 20° $\underline{1.42}$ (174, 249, 435).

2-Fluoronaphthalene, B 20° $\underline{1.56}$ (131); B 19° $\underline{1.49}$ (174, 249, 435).

$C_{10}H_7I$ 1-Iodonaphthalene, B 20° $\underline{1.43}$ (131).

2-Iodonaphthalene, B 20° $\underline{1.56}$ (131).

$C_{10}H_7N$ p-Tolylpropiolnitrile, B 25° $\underline{4.90}$ (529).

$C_{10}H_7NO_2$ 1-Nitronaphthalene, B $\underline{3.62}$ (79); B 20° $\underline{3.88}$ (174, 249, 435).

2-Nitronaphthalene, B 25° $\underline{4.4}$ (733); B 25° $\underline{4.36}$ (747).

α-Nitroso-β-naphthol, B or Hx $\underline{4.39}$ (237).

β-Nitroso-α-naphthol, B or Hx $\underline{4.36}$ (237).

$C_{10}H_8$ Naphthalene, 25° Hx $\underline{0.72}$, CD $\underline{0.69}$ (63); B 20° $\underline{0}$ (131); B 25° $\underline{0}$ (201); Hp 20° $\underline{0}$ (220).

$C_{10}H_8BrN$ 2-Bromo-6-naphthylamine, B 25° $\underline{3.31}$ (733).

$C_{10}H_8Br_2O_2$ Methyl α,β-dibromocinnamate, MP 100°, B 21° $\underline{2.68}$ (461); MP 134°, B 21° $\underline{1.98}$ (461).

$C_{10}H_8N_2$ 2,2'-Bipyridyl, B 17-67° < $\underline{0.68}$ (416).

4,4'-Bipyridyl, B 17-67° < $\underline{0.55}$ (416).

$C_{10}H_8N_2O_2$ 1,2- or β-Naphthoquinone dioxime, D 25° $\underline{3.80}$ (607).

1-Nitro-2-naphthylamine, B 40° 4.47 (624, 625, 747); B 25° 4.63 (733).
1-Nitro-4-naphthylamine, B 25° 6.38 (733).
1-Nitro-5-naphthylamine, B 25° 4.96 (733).
2-Nitro-1-naphthylamine, B 40° 4.89 (624, 625, 747); B 25° 4.92 (733).
2-Nitro-6-naphthylamine, B 25° 5.14 (733).
3-Nitro-1-naphthylamine, B 25° 5.14 (747).
4-Nitro-1-naphthylamine, D 25° 6.97 (747).
4-Nitro-2-naphthylamine, B 25° 4.62 (747).
5-Nitro-1-naphthylamine, B 25° 5.22 (624, 747).
5-Nitro-2-naphthylamine, B 25° 5.03 (624, 747).
6-Nitro-2-naphthylamine, D 25° 7.10 (747).
8-Nitro-1-naphthylamine, B 25° 3.12 (747).
8-Nitro-2-naphthylamine, B 25° 4.47 (747).

$C_{10}H_8O$ 1- or α-Naphthol, B 21° 1.0 (253); E 1.8 (298); B 20-60° 1.40 to 1.43, E 26° 1.91 (300); E 20° 1.9, 26° 1.91 (358).
2- or β-Naphthol, B 18° 1.3 (253); B 25° 1.53 (300).
Phenyl-acetylacetylene, B 25° 3.23 (533).

$C_{10}H_9BrO_2$ Methyl β-bromocinnamate, B 22° 2.63 (461).

$C_{10}H_9N$ 2-Methylquinoline (quinaldine), B 10-40° 1.86 ± .01 (382).
1- or α-Naphthylamine, B 23° 1.44 (463); 20° B 1.53, T 1.48, CH 1.47 (585); B 25° 1.44 (733); B 25° 1.49 (747).
2- or β-Naphthylamine, B 23° 1.74 (463); 20° B 1.76, T 1.73 (585); B 25° 1.84 (733); 25° B 1.77, D 2.10 (747).

$C_{10}H_9NO$ α-Methyl-γ-phenylisoxazole, B 3.18 (704).
γ-Methyl-α-phenylisoxazole, B 3.21 (704).

$C_{10}H_9NO_2S$ α-Naphthalenesulfonamide, D 5.14 (718, 719); D 25° 5.12 (849).
β-Naphthalenesulfonamide, D 5.36 (718, 719); D 25° 5.27 (849).

$C_{10}H_{10}$ 1,2-Dihydronaphthalene, B 25° 1.4 (378).
1,4-Dihydronaphthalene, Hx 25° 1.4 (378).
p-Ethylphenylacetylene, S 1.05 (373).

$C_{10}H_{10}Br_2$ 6,7-Dibromotetralin, B 25° 2.81 (507).
$C_{10}H_{10}N_2$ 2,3-Dimethylquinoxaline, B 0.8 and 0 (273); B< 0.3 (335).
$C_{10}H_{10}N_2O$ 6-Methoxy-8-aminoquinoline, B 1.89 (742).
$C_{10}H_{10}N_2O_2$ Methyl-p-anisyl furazan, B 25° 4.68 (666).
$C_{10}H_{10}N_2O_3$ Methyl-p-anisyl furazan peroxide, B 25° 4.68 (666).
Methyl-p-anisyl furoxan, B 25° 5.55 (666).

$C_{10}H_{10}O$ Benzalacetone, B 3.31 (119).
$C_{10}H_{10}O_2$ p-Diacetylbenzene, B 25° 2.7₁ (822).
Methyl cinnamate, D 25° 1.93 (589).

$C_{10}H_{10}O_4$ Dimethyl o-phthalate, B 2.3 (42); B 25° 2.8 (95).
Dimethyl p-phthalate (dimethyl terephthalate), B 2.2 (42); B 25° 2.2 (61, 95).
Hydroquinone diacetate, B 25° 2.2 (58, 60, 61, 95).
cyclic-Phenylene dioxene ether, S 1.39 (354).

$C_{10}H_{11}Br$ 7-Bromo-1,2,3,4-tetrahydronaphthalene (7-bromotetralin), B 25° 2.23 (507).

$C_{10}H_{11}BrO_2$ Methyl p-bromo-β-phenylpropionate, B 20° 2.39 (461).

$C_{10}H_{11}N$ 1,2-Dimethylindole, B 25° 2.52 (797).
1,3-Dimethylindole, B 25° 2.04 (797).
2,3-Dimethylindole, B 25° 2.39 (797).

$C_{10}H_{11}NO_2$ 1-Nitro-1,2,3,4-tetrahydronaphthalene (1-nitrotetralin), B 25° 3.98 (747).
2-Nitrotetralin, B 25° 4.81 (747).

$C_{10}H_{12}$ o,α-Dimethylstyrene, B 26° 0.8 (464).
4-Ethylphenylethylene (p-ethylstyrene), B 25° 0.61 (439).

1,2,3,4-Tetrahydronaphthalene (tetralin), B 25° 1.66 ± .04 (314); B 10-40° 0.5 (446); B 25° 0.52 (507); S 20° 0.6 (559).

$C_{10}H_{12}BrNO_2$ — Bromonitrodurene, B 25° 2.36 (519).

$C_{10}H_{12}Cl_2$ — Dichlorodurene, B 25° ca. 0.25 (701).

Dichloroprehnitene, B 25° 2.93 (701).

$C_{10}H_{12}N_2O_4$ — Dinitrodurene, B 25° 0.60 (519).

Dinitroprehnitene, B 25° 6.86 (701).

$C_{10}H_{12}O$ — β- or ω-Ethoxystyrene, B 18° 1.68 (461).

o-Methoxy-α-methylstyrene, B 27° 1.48 (464).

p-Methoxy-α-methylstyrene, B 26° 1.39 (464).

$C_{10}H_{12}O_2$ — Ethyl phenylacetate, S 30° 1.82 (738).

$C_{10}H_{13}Br$ — 2-Bromo-p-cymene, B 25° 1.76 (428).

3-Bromo-p-cymene, B 25° 1.71 (428).

Bromodurene, B 25° 1.55 (519).

$C_{10}H_{13}Cl$ — 2-Chloro-p-cymene, B 25° 1.64 (428).

3-Chloro-p-cymene, B 25° 1.66 (428).

$C_{10}H_{13}NO_2$ — o-Benzbetaine, B-ethyl alc. 25° 6.05 (414).

m-Benzbetaine, B-ethyl alc. 25° 7.36 (414).

p-tert-Butylnitrobenzene, B 25° 4.61 (428).

Methyl N-dimethylanthranilate, B 25° 2.05 (414).

Nitrodurene, B 25° 3.39 (519).

$C_{10}H_{13}NO_3$ — Nitrodurenol, B 25° 4.08 (648).

$C_{10}H_{14}$ — tert-Butylbenzene, B 25° 0.53 (428); 456-477°K 0.70 (627).

p-Cymene, B or CT 0, 20 and 45° 0.15 (428).

1,4-Diethylbenzene, B 25° 0.2₄ (94).

$C_{10}H_{14}BrN$ — Bromoaminodurene, B 25° 2.75 (519).

$C_{10}H_{14}N_2O$ — p-Nitroso-N-diethylaniline, B 25° 7.18, CT 25° 6.87, 45° 6.90 (240); B 7.15 (678).

$C_{10}H_{14}N_2O_2$ — Nitroaminodurene, B 25° 4.98 (519).

$C_{10}H_{14}O$ — Carvacrol, B 25° 1.54 (428).

Carvone, B 15° ca. 2.77 (100).

Durenol, B 25° 1.68 (648).

Thymol, B 25° 1.54 (428); B 18° 1.60 (799).

$C_{10}H_{14}O_2$ — o-Diethoxybenzene (catechol diethyl ether), B 25° 1.3₇ (94); Dec 25° 0.8, (159).

m-Diethoxybenzene (resorcinol diethyl ether), B 25° 1.7₀ (94).

p-Diethoxybenzene (hydroquinone diethyl ether), B 25° 1.7 (58, 60, 61, 95); B 18° 1.77 (78); B 25° 1.7₆, CH 1.7₄, CT 1.7₂, CD 1.7₄ (94); B 20° 1.72, 40° 1.92, 60° 2.00 (98); B 18° 1.72 40° 1.75, 60° 1.78 (116, 120).

$C_{10}H_{15}N$ — Aminodurene, B 25° 1.39 (519).

N-Diethylaniline, B 18° 1.65 (799).

$C_{10}H_{16}$ — Limonene, B 15° < 0.5 (100); d-, B 25° 1.56 (454).

Pinene, d-, B 30° 2.67 (454); l-, S 20° 0.8 (559); dl-, S 20° 0.6 (559); B 30° 0.36 (779).

$C_{10}H_{16}N_2$ — Decanedinitrile, B 25° 4.47, 75° 4.59 (569); B 25° 4.50, 75° 4.66 (669).

N,N'-Tetramethyl-p-phenylenediamine, B 25° 1.23 (93).

$C_{10}H_{16}O$ — Camphor, S 22° 2.95 ± .03 (108); B 22° 2.94 ± .03 (147); d-, B or Hx 3.05 (237).

Fenchone, S 22° 2.92 ± .03 (108); B 22° 2.90 ± .03 (147).

$C_{10}H_{16}O_2$ — 2,2-Dimethylmethone, B 22° 3.20 (571, 622).

3-Ethoxymethone, B 20° 3.75 (571, 622).

$C_{10}H_{16}O_4$ — Cyclohexane-1,4-diol diacetate (quinitol diacetate), B 18° 1.46 (234).

1,4-Diacetoxy-2,3-dimethyl-2-butene, cis-, B 25° 2.58 (845); trans-, B 25° 1.90 (845).

Diethyl cyclobutane-1,1-dicarboxylate, B 25° 2.22 (290); B 2.22 (773).

Dimethyl hexahydroisophthalate, cis-, S 20° 2.30 (663); trans-, S 20° 2.09 (663).

$C_{10}H_{18}$ **Decahydronaphthalene** (decalin), Lqd. 20-175° 0 (44); Lqd. 25-142° 0 (322); cis-, B or CT 25° 0 (314); B 20-100° ca. 0 (868); trans-, B 25° 0 (314); B 20-100° ca. 0 (868).

Di-n-butylacetylene, Hp 25° 0 (396).

$C_{10}H_{18}N_2O_4$ **Methyl** 1-nitrosoisopropyl ketone dimer, 25° B 3.32, D 3.38 (730).

$C_{10}H_{18}O$ **Borneol**, E 20° 1.7 (298, 358); dl-, B or Hx 1.65 (237); 1-, S 22° 1.56 ± .02 (109); B 7° 1.65 ± .03 (353).

Menthone, B 15° 2.77 (99, 100); S 22° 2.80 ± .03 (108); S 2.80 (117); B 22° 2.82 ± .03 (147).

$C_{10}H_{18}O_2$ β,γ-Dimethyl-ε-octolactone, B 25° 4.33 (496).

$C_{10}H_{18}O_4$ Diethyl adipate, B 25° 2.40, 50° 2.42 (185); B 25 and 50° 2.40 (315).

Diethyl isopropylmalonate, B 2.40 (773).

$C_{10}H_{18}O_6$ Diethyl 1,2-dimethoxysuccinate, dl-, B 25° 3.74 (202); meso-, B 25° 3.34 (202).

$C_{10}H_{20}$ tert-Butylcyclohexane, 411-456°K 0 (627).

$C_{10}H_{20}Br_2$ Decamethylene dibromide, 197°K 2.75, 215° 2.75, 221° 2.73, 229° 2.69 (184); B 25° 2.54, 50° 2.56 (259).

$C_{10}H_{20}O$ Menthol, 1-, S 22° 1.57 ± .03 (109); B 7° 1.62 ± .03 (353).

1,2,4,5-Tetramethylcyclohexanol, B 18° 1.94 (234).

$C_{10}H_{20}O_2$ 2-Methyl-2-amyl-1,3-dioxane, B 25° 1.90 (560).

sec-n-Octyl acetate (2-acetoxy-n-octane), B 25° 1.9_3 (127).

$C_{10}H_{21}Br$ n-Decyl bromide, B 1.90 (721).

$C_{10}H_{22}$ n-Decane, Lqd. -30° to 170° ca. 0 (110).

$C_{10}H_{22}O$ n-Amyl ether, Lqd. 30-40° 1.04 (815).

n-Decyl alcohol, S 1.63 (74); B 25° 1.61 (82).

Isoamyl ether, B 20-50° 0.97 (40); B 25° 1.23 (836).

Tri-n-propylcarbinol, B 18° 1.65 (234).

$C_{10}H_{22}O_2$ 1,10-Decanediol (decamethylene glycol), D 25° 2.52, 50° 2.54 (186); D 15° 2.36 ± .03 (707).

$C_{10}H_{22}S$ n-Amyl sulfide, B 25° 1.58, 50° 1.59 (326).

$C_{11}H_6ClN$ 8-Chloro-1-naphthonitrile, B 19° 5.70 (462).

$C_{11}H_7NO$ α-Naphthyl isocyanate, B 20° 2.30 (467).

β-Naphthyl isocyanate, B 20° 2.34 (467).

$C_{11}H_8O_2$ Cyclopentadiene-p-benzoquinone, B or CT 25 and 45° 0.69 (427).

$C_{11}H_9NO_4$ Ethyl p-nitrophenylpropiolate, B 22° 3.54 (461).

$C_{11}H_{10}O_2$ Ethyl phenylpropiolate, B 21° 2.19 (461).

$C_{11}H_{11}BrO_2$ Ethyl α-bromocinnamate, MP 126°, B 19° 2.80 (461); MP 131°, B 22° 2.25 (461).

Ethyl β-bromocinnamate, B 19° 1.81 (461).

$C_{11}H_{11}NO_4$ Ethyl p-nitrocinnamate, B 21° 3.50 (461).

$C_{11}H_{11}N_3O_2S$ 2-Sulfanilamidopyridine (sulfapyridine), D 25° 6.8 ± .2 (759).

$C_{11}H_{12}$ p-Isopropylphenylacetylene, S 1.12 (373).

$C_{11}H_{12}N_2O$ 1,5-Dimethyl-2-phenyl-3-pyrazole (antipyrine), B 25° 5.48 (787).

$C_{11}H_{12}N_2S$ 1,5-Dimethyl-2-phenyl-3-pyrazolthione (thiopyrine), B 25° 7.33 (787).

$C_{11}H_{12}N_4O_2S$ 2-Sulfanilamido-4-methylpyrimidine, D 25° 7.2 ± .2 (759).

$C_{11}H_{12}O_2$ Ethyl allocinnamate (cis), B 20° 1.77 (461); ethyl cinnamate (trans), B 20° 1.84 (461); B 23° 1.84 (631); Lqd. 35-40° 2.10 (815).

$C_{11}H_{12}O_4$	3,5-Diacetyl-2,6-dimethyl-γ-pyrone, S 25° 4.0_6 (552).
$C_{11}H_{12}O_5$	Ethyl cinnamate ozonide (trans), B 23° 2.00 (631).
$C_{11}H_{13}BrN_2O_2$	4'-Bromo-2'-nitro-1-phenylpiperidine, B 25° 4.29 (492).
$C_{11}H_{13}N$	2-Methyl-3-ethylindole, B 25° 2.41 (797).
	3-Methyl-2-ethylindole, B 25° 2.36 (797).
$C_{11}H_{13}NO$	p-Dimethylaminocinnamic aldehyde, B 21° 5.4 (708).
$C_{11}H_{13}NO_4$	Ethyl p-nitro-β-phenylpropionate, B 20° 4.58 (461).
$C_{11}H_{14}O_2$	Ethyl phenyl-methylacetate, S 30° 1.818 (738).
	Ethyl β-phenylpropionate, B 21° 1.75 (461).
$C_{11}H_{15}Cl$	Pentamethylchlorobenzene, B 25° 1.85 (701).
$C_{11}H_{16}$	p-tert-Butyltoluene, B 25° 0.35 (428); 477°K 0.39 (627).
$C_{11}H_{16}N_2O$	N,N-Diethyl-N'-phenylurea, D 17° $3.2 \pm .1$ (578).
$C_{11}H_{17}N$	N-Dimethylmesidine, B 25° 1.03 (648).
	Pentamethylaniline, B 25° 1.10 (519).
$C_{11}H_{18}O_2$	Bornyl formate, 1-, S 22° $2.04 \pm .01$ (109).
	3-Isopropoxymethone, B 20° 3.78 (571, 622).
$C_{11}H_{18}O_4$	Diethyl cyclopentane-1,1-dicarboxylate, B 25° 2.14 (290); B 2.14 (773).
$C_{11}H_{20}BrN$	11-Bromoundecanenitrile, B 25° 3.92, 75° 3.95 (669).
$C_{11}H_{20}O_2$	Menthyl formate, S 22° $2.06 \pm .03$ (109).
$C_{11}H_{20}O_4$	Erythritol diisopropylidene ether (diacetonepentaerythritol), B 22° $2.26 \pm .07$ (128).
	3,4-Diacetoxyheptane, dl-threo-, Lqd. 25° 2.18 (710); dl-erythro-, Lqd. 25° 2.68 (710).
$C_{11}H_{20}O_4$	Diethyl diethylmalonate, B 2.10 (773).
$C_{11}H_{22}N_2O_2$	Methyl-n-octylglyoxime, D 20° 1.485 (366).
$C_{11}H_{22}O$	Methyl nonyl ketone, B 15° 2.69 (66).
$C_{11}H_{24}$	n-Undecane, Lqd. -11 to 190° ca. 0 (110).
$C_{11}H_{24}O$	n-Undecyl alcohol, B 25° 1.66 (80).
$C_{11}H_{24}O_2$	Bis(2,2-dimethylpropoxy)methane (dineopentoxymethane), B 25° 0.88 (437).
	2,2-Diethoxyheptane, B 25° 0.90 (437).
$C_{12}H_6Br_2N_2O_4$	4,4'-Dibromo-2,2'-dinitrobiphenyl, B 25° 4.92 (601).
$C_{12}H_6Cl_2N_2O_4$	4,4'-Dichloro-2,2'-dinitrobiphenyl, B 25° 4.90 (601).
$C_{12}H_6Cl_2O_2$	2,6-Dichlorodi-o-phenylene dioxide, B 25° 0.62 (724, 725, 745).
$C_{12}H_6Cl_2S_2$	2,6-Dichlorodi-o-phenylene disulfide (2,6-dichlorothianthrene), B 25° 1.37 (576).
$C_{12}H_6N_4O_8$	2,2',4,4'-Tetranitrobiphenyl, B 25° 4.38 (601).
$C_{12}H_6O_2$	Acenaphthenequinone (1,2-acenaphthenedione), B 25° 6.0_8 (634).
$C_{12}H_8Br_2$	p,p'-Dibromobiphenyl, B 25° 0 (34).
$C_{12}H_8Br_2N_2$	p,p'-Dibromoazobenzene, CD 21° < 1 (101).
$C_{12}H_8Br_2O$	p-Bromophenyl ether, B 18° 0.59 (216); B 25° 0.60, 50° 0.62, Hp 25 and 50° 0.62 (260); Hx 25° 0.65 (453); 517°K $0.86 \pm .05$ (584).
$C_{12}H_8Br_2S$	p-Bromophenyl sulfide, B 25° 0.65, 50° 0.67 (260).
$C_{12}H_8Br_2Se_2$	p-Bromophenyl diselenide, B 25° 0.70 (860).
$C_{12}H_8Cl_2$	o,o'-Dichlorobiphenyl, B 25° $1.72 \pm .02$ (68); 25° B 1.7_7, CT 1.7_1 (327); B 25° 1.91 (473).
	m,m'-Dichlorobiphenyl, 25° B 1.6_8, CT 1.7_1 (327); B 25° 1.80 (473).
	p,p'-Dichlorobiphenyl, B 25° 0.33 (34); B 25° 0 (61, 65, 95); 25° B ca. 0, CB 0.82, NB 2.00 (491).
$C_{12}H_8Cl_2OS$	p-Chlorophenyl sulfoxide, B 22° 2.7 (218); B 25° 2.57 (296).
$C_{12}H_8Cl_2O_4S_4$	Diphenyldisulfide-4,4'-disulfonyl chloride, B 25° 4.64 (848).
$C_{12}H_8Cl_2S$	p-Chlorophenyl sulfide, B 22° 0.89 (216); B 25° 0.58 (296).
$C_{12}H_8Cl_2Se$	p-Chlorophenyl selenide, B 25° 0.77 (860).
$C_{12}H_8F_2$	p,p'-Difluorobiphenyl, B 25° 0.35 (34).

$C_{12}H_8I_2O$ o-Iodophenyl ether, B 27° 2.72, 50° 2.78, Hx 2.70 (647, 726).

$C_{12}H_8N_2$ Phenazine, B 15° 0 (211); B 25° ca. 0 (582); 0 (598).

$C_{12}H_8N_2O_4$ o,o'-Dinitrobiphenyl, B 25° 5.12 ± .05 (68); B 25° 5.19 (601).
p,p'-Dinitrobiphenyl, B 25° 0 (61, 65, 95); B 25° 1.0_2 (604).

$C_{12}H_8N_2O_4S_2$ p-Nitrophenyl disulfide, B 20° 3.56 (216).

$C_{12}H_8N_2O_5$ o-Nitrophenyl ether, B 27° 6.64, 50° 6.57, D 27° 6.72 (647, 672).
p-Nitrophenyl ether, B 25° 2.79, 50° 2.80 (260); B 25° 2.61 (296).

$C_{12}H_8N_3O_5$ p,p'-Dinitrodiphenyl nitroxide, S 2.7 ± .5 (277).

$C_{12}H_8O$ Di-o-phenylene oxide (dibenzofuran), B 25° 0.88 (68).

$C_{12}H_8OS$ Phenothioxine (phenoxthine), 25° B or Hx 1.09 (724, 726).

$C_{12}H_8O_2$ Di-o-phenylene dioxide, 20° B, CH, or CT 0 (338); 25° B 0.64, Hx 0.57 (724, 725).

$C_{12}H_8O_2S_2$ Thianthrene dioxide ("thianthrene disulfoxide, MP 246°"), B 19° 4.2 (218).
Thianthrene monosulfone ("thianthrene disulfoxide, MP 279°"), B 19° 1.7 (218).

$C_{12}H_8S$ Di-o-phenylene sulfide (dibenzothiophene), B 24° ca. 0 (216).

$C_{12}H_8S_2$ Di-o-phenylene disulfide (thianthrene), B 22° 1.7 (218); B 25 and 50° 1.41 (326); CT 25° 1.54, CD 24° 1.47 (339); B 25° 1.57 (582).

$C_{12}H_8Se_2$ Selenanthrene, B 25° 1.41 (582).

$C_{12}H_9Br$ 4-Bromobiphenyl, B 25° 1.64 (492).

$C_{12}H_9BrN_2$ p-Bromoazobenzene, B 20° 1.42 (101).

$C_{12}H_9BrN_2O$ α-4-Bromoazoxybenzene, B 2.60_5 (294).
β-4-Bromoazoxybenzene, B 0.90 (294).

$C_{12}H_9BrO$ p-Bromophenyl phenyl ether, B 19° 1.76 (216); B 25° 1.56, 50° 1.58, Hp 25° 1.65, 50° 1.67 (260); Hx 25° 1.75 (453); 516°K 1.90 ± .02 (584).

$C_{12}H_9Br_2N_3$ 4,4'-Dibromodiazoaminobenzene, B 25° 1.88 ± .04 (553).

$C_{12}H_9Cl$ o-Chlorobiphenyl, 25° B 1.3_0, CT 1.4_2 (327); B 1.44 (369); B 25° 1.45 (473).
m-Chlorobiphenyl, 25° B 1.7_9, CT 1.60 (327); B 25° 1.64 (473).
p-Chlorobiphenyl, B 25° 1.5_3 (327); B 1.63 (369); B 25° 1.63 (473); B 25° 1.56 (492).

$C_{12}H_9ClN_2$ p-Chloroazobenzene, B 21° 1.55 (101).

$C_{12}H_9ClOS$ p-Chlorophenyl phenyl sulfoxide, B 23° 3.94 (218); B 25° 3.47 (296).

$C_{12}H_9ClO_2S$ p-Chlorophenyl phenyl sulfone, B 23° 4.42 (218).

$C_{12}H_9ClS$ p-Chlorophenyl phenyl sulfide, B 21° 1.76 (103); B 21° 1.70 (216); B 25° 1.51 to 1.54 (296).

$C_{12}H_9Cl_2N_3$ 4,4'-Dichlorodiazoaminobenzene, B 25° 1.94 ± .04 (553).

$C_{12}H_9F$ 4-Fluorobiphenyl, B 25° 1.50 (492).

$C_{12}H_9N$ Carbazole, B 20° 2.09 (468).

$C_{12}H_9NO_2$ 2-Nitrobiphenyl, B 3.79 (369); B 25° 3.80 (492).
3-Nitrobiphenyl, B 3.90 (369).
4-Nitrobiphenyl, B 4.28 (369); B 25° 4.17 (492).

$C_{12}H_9NO_3$ p-Nitrophenyl phenyl ether, B 19° 4.20 (216); 25° B 4.26 to 4.29, CT 4.30 to 4.34 (296); 499-516°K 4.44 ± .18 (584).

$C_{12}H_{10}$ Acenaphthene, B 20° 1.57, 25° 1.5_0, 30° 1.6_0 (855).
Biphenyl (diphenyl), B 25° 0 (34, 61, 95); Lqd. 75-155° 0 (44); Hp 20° 0 (220); 25° B 0, CB 0.96, NB 1.88 (491).

$C_{12}H_{10}BrN_3$ 4-Bromodiazoaminobenzene (1-phenyl-3-p-bromophenyltriazene), B 25° 2.00 ± .04 (553).

$C_{12}H_{10}Br_2Se$ Diphenylselenium dibromide, B 25° 3.40 (784).

$C_{12}H_{10}Cl_2Se$ Diphenylselenium dichloride, B 50° 3.47 (702); B 25° 3.21 (784).

$C_{12}H_{10}NO$ — Diphenyl nitroxide, S $\underline{2.3} \pm .2$ (277).

$C_{12}H_{10}N_2$ — Azobenzene, B 24° $\underline{0}$ (101); D 15° $\underline{0}$ (463); cis-, B 25° $\underline{3.0}$ (592, 645); trans-, B 25° $\underline{0}$ (592, 645).

$C_{12}H_{10}N_2O$ — Azoxybenzene, cis-, $\underline{4.67}_8$ (294); trans-, B $\underline{1.70}_5$ (294).
o-Hydroxyazobenzene, B 15° $\underline{1.31}$ (463).
p-Hydroxyazobenzene, 15° B $\underline{1.62}$, D $\underline{2.04}$ (463).
N-Nitrosodiphenylamine, B 20° $\underline{3.39}$ (280).

$C_{12}H_{10}N_2O_2$ — p-Azophenol, cis-, D 25° $\underline{2.69}$ (638); trans-, D 25° $\underline{2.60}$ (638).
4-Nitro-4'-aminobiphenyl, B 25° $\underline{6.46}$ (492).
o-Nitrophenyl phenylamine, B 20° $\underline{4.13}$ (595).
p-Nitrophenyl-phenylamine, B 20° $\underline{5.82}$ (595).

$C_{12}H_{10}N_4O_2$ — 4-Nitrodiazoaminobenzene (1-phenyl-3-p-nitrophenyltriazene),
B 25° $\underline{4.77} \pm .06$ (553).

$C_{12}H_{10}O$ — Phenyl ether, Lqd. 28-50° $\underline{0.98}$, B 20-50° $\underline{1.02}$ (40); 373°K $\underline{1.0}$
(41); B 18° $\underline{1.13}$ (216); Hx 25° $\underline{1.18}$ (453); 486°K $\underline{1.35}$ (538);
445-483°K $\underline{1.14} \pm .02$ (584); Lqd. 30-40° $\underline{1.14}$ (815).

$C_{12}H_{10}OS$ — Phenyl sulfoxide, B 23° $\underline{4.08}$ (103); B 25° $\underline{4.17}$ (160); B 25°
$\underline{4.00}$ (296).

$C_{12}H_{10}OSe$ — Diphenylseleno oxide, B 25° $\underline{4.44}$ (786).

$C_{12}H_{10}O_2S$ — Phenyl sulfone, B 25° $\underline{5.05}$ (160); B 23° $\underline{5.14}$ (218).

$C_{12}H_{10}O_4S_2$ — Phenyl disulfone, B 25° $\underline{3.93}$ (770).

$C_{12}H_{10}S$ — Phenyl sulfide, B 21° $\underline{1.47}$ (103); B 25° $\underline{1.565}$ (160); B 25° $\underline{1.50}$
(296).

$C_{12}H_{10}S_2$ — Phenyl disulfide, B 24° $\underline{1.81}$ (216).

$C_{12}H_{10}Se$ — Phenyl selenide, B 20° $\underline{1.38}$ (103).

$C_{12}H_{10}Se_2$ — Phenyl diselenide, B 25° $\underline{1.67}$ (860).

$C_{12}H_{11}N$ — 2-Aminobiphenyl, B $\underline{1.42}$ (369).
4-Aminobiphenyl (xenylamine), B $\underline{1.73}$ (369); B 25° $\underline{1.76}$ (492);
25° B $\underline{1.74}$, D $\underline{5.09}$ (732).
Diphenylamine, Solid 20° $\underline{1.3}$ (40).

$C_{12}H_{11}NO_2S$ — p-Phenylbenzenesulfonamide, D 25° $\underline{5.20}$ (732).

$C_{12}H_{11}N_3$ — p-Aminoazobenzene, B 15° $\underline{2.71}$ (463).
Diazoaminobenzene, B 25° $\underline{0.9} \pm .02$ (553).

$C_{12}H_{12}N_2$ — 2,2'-Diaminobiphenyl, B 25° $\underline{2.0} \pm .002$ (68).
4,4'-Diaminobiphenyl (benzidine), B 25° $\underline{1.43}$ (34); B 25° $\underline{1.3}$
(61, 65, 95).
1,1-Diphenylhydrazine, B 20° $\underline{1.87}$ (275).
1,2-Diphenylhydrazine (hydrazobenzene), B 18° $\underline{1.53}$ (275); B
20° $\underline{1.66}$ (280).

$C_{12}H_{12}N_2O_4S_4$ — Phenyldisulfide-4,4'-disulfonamide, D 25° $\underline{6.37}$ (849).

$C_{12}H_{12}O_6$ — Phloroglucinol triacetate, B 25° $\underline{2.4}$ (58, 61).

$C_{12}H_{14}O_4$ — Diethyl o-phthalate, B $\underline{2.4}$ (42); 25° B $\underline{2.7}$, D $\underline{2.8}$ (144); 25°
Dec $\underline{2.69}$, tetrachloroethylene $\underline{2.68}$ (159).
Diethyl terephthalate, B $\underline{2.3}$ (42).

$C_{12}H_{15}N$ — 2-tert-Butylindole, B 25° $\underline{2.44}$ (797).
1,2-Dimethyl-3-ethylindole, B 25° $\underline{2.41}$ (797).
1,3-Dimethyl-2-ethylindole, B 25° $\underline{2.42}$ (797).

$C_{12}H_{15}NO$ — N-Dimethyl-p-aminobenzylideneacetone, B 24° $\underline{5.3}$ (708).

$C_{12}H_{15}N_3O_6$ — Trinitro-5-tert-butyl-m-xylene, B 25 and 45° $\underline{1.14}$, CT 25°
$\underline{1.14}$ (426).
Trinitro-2,4,6-triethylbenzene, B 25° $\underline{0.8}$ (479).

$C_{12}H_{16}N_2S$ — N-Phenyl-N-ethyl-N'-allylthiourea, B 18° $\underline{5.1}$ (799).

$C_{12}H_{16}O_2$ — Isoamyl benzoate, CT $\underline{2.2}$ (208).

$C_{12}H_{17}NO_3$ — Nitroethoxydurene, B 25° $\underline{3.69}$ (648).

$C_{12}H_{18}$ — 5-tert-Butyl-m-xylene, B 25° $\underline{0.25}$ (428).
Hexamethylbenzene (mellitene), B 20° $\underline{0.1}$ (192).
1,3,5-Triethylbenzene, B 25° $\underline{0.1}$ (58); B 25° < $\underline{0.2}$ (61).

$C_{12}H_{18}N_2O_2$	Nitro-N-dimethylaminodurene, B 25° 4.11 (648).
$C_{12}H_{20}N_2$	n-Dodecanedinitrile, B 25° 4.95, 50° 5.00 (669).
$C_{12}H_{20}O_2$	Bornyl acetate, 1-, S 22° 1.87 ± .02 (109).
$C_{12}H_{20}O_4$	Diethyl cyclohexane-1,1-dicarboxylate, B 2.23 (290, 773).
	Diethyl hexahydroisophthalate, cis-, S 20° 2.35 (663); trans-, S 20° 2.13 (663).
$C_{12}H_{22}$	Di-n-amylacetylene, B 25° 0 (396).
$C_{12}H_{22}O_2$	Menthyl acetate, S 22° 1.83 ± .03 (109).
$C_{12}H_{22}O_{11}$	Sucrose, 20° pyridine 2.8, butylamine or diethylamine 3.4 (364).
$C_{12}H_{24}O_2$	Lauric acid, B 25° 0.75 (698).
$C_{12}H_{25}I$	1-Iodododecane (lauryl iodide), CT 20° 1.85 (750).
$C_{12}H_{26}$	n-Dodecane, Lqd. -10 to 210° ca. 0 (110).
$C_{12}H_{26}O$	n-Dodecyl alcohol, B 25° 1.62 (74, 80).
$C_{12}H_{28}BrN$	Tri-n-butylammonium bromide, B 25° 7.61 (472).
$C_{12}H_{28}ClN$	Tri-n-butylammonium chloride, B 25° 7.17 (472).
$C_{12}H_{28}IN$	Tri-n-butylammonium iodide, B 25° 8.09 (472).
$C_{13}H_6Br_2O$	2,7-Dibromo-9-fluorenone, B 25° 4.44 (541).
$C_{13}H_6Br_2O_2$	β- or 2,7-Dibromoxanthone, B 25° 4.10 (551).
$C_{13}H_6N_2O_5$	2,5-Dinitro-9-fluorenone, B 25° ca. 6 (542).
	2,7-Dinitro-9-fluorenone, B 25° 4.80 (542).
$C_{13}H_6N_2O_6$	α- or 2,4-Dinitroxanthone, B 25° 2.98 (551).
	β- or 2,7-Dinitroxanthone, B 25° 5.72 (551).
$C_{13}H_7NO_3$	2-Nitro-9-fluorenone, B 25° 6.04 (542).
$C_{13}H_8Br_2$	2,7-Dibromofluorene, B 15° 0 (209); B 25° 0.22 (542).
$C_{13}H_8Br_2O$	p,p'-Dibromobenzophenone, B 22° 1.69 (211).
$C_{13}H_8Cl_2O$	p,p'-Dichlorobenzophenone, B 13° 1.64 (211); B 1.57 ± .03 (293); Hx 25° 1.70 (453).
	9,9-Dichlorofluorene, B 20° 1.85 (209).
$C_{13}H_8Cl_4$	Di-p-chlorophenyldichloromethane, B 17° 0.48 (210).
$C_{13}H_8N_2O_4$	2,5-Dinitrofluorene, B 25° 7.06 (542).
	2,7-Dinitrofluorene, B 25° ca. 0.17 (542); S 25° ca. 2.39 (821).
$C_{13}H_8O$	9-Fluorenone, B 18° 3.29 (209); B 25° 3.35 (542).
$C_{13}H_8OS$	Thioxanthone, D 17° 5.4 (709).
	Xanthione, B 28° 5.4 (709).
$C_{13}H_8O_2$	Xanthone, B 14° 3.07 (463); B 10° 2.91, 20° 2.93, 30° 2.94, 40° 2.95 (506); B 25° 3.11 (551); B 15° 2.92 ± .02 (573).
$C_{13}H_8S_2$	Thioxanthione, D 31° 5.2 (709).
$C_{13}H_9BrO$	p-Bromobenzophenone, B 20° 2.75 (211).
$C_{13}H_9Cl$	9-Chlorofluorene, B 14° 1.76 (209).
$C_{13}H_9ClO$	p-Chlorobenzophenone, B 13° 2.70 (211); S 2.64 (354); Hx 25° 2.71 (453).
$C_{13}H_9Cl_2N$	p-Chlorobenzophenonechloroimine, syn- or β-, B 25° 2.67 ± .05 (667); anti- or α-, B 25° 2.47 ± .06 (667).
	p-Chlorobenzylidene-p-chloroaniline, B 25° 1.56 (588).
$C_{13}H_9N$	Acridine, B 14° 1.95 (211).
	o-Phenylbenzonitrile, B 18° 3.81 (461).
$C_{13}H_9NO_2$	2-Nitrofluorene, D 11° 5.44 (209); D 25° 4.1 (821).
$C_{13}H_9N_3O_4$	p-Nitrobenzylidene-p-nitroaniline, B 25° 3.56 (729).
$C_{13}H_{10}$	Fluorene, B 14° 0.28 (209); B 25° 0.82 (542); B 0.53, D 0.65 (806).
$C_{13}H_{10}BrN$	Benzophenonebromoimine, B 25° 2.83 ± .03 (667).
$C_{13}H_{10}BrNO$	5-Bromosalicylideneaniline, B 25° 1.19 (588).
$C_{13}H_{10}Br_2$	Di-p-bromophenylmethane, B 17° 1.79 (210); B 25° 1.85 to 1.88 (296).
$C_{13}H_{10}ClN$	Benzophenonechloroimine, B 25° 2.96 ± .04 (667).
	p-Chlorobenzylideneaniline, B 25° 1.77 (588).

$C_{13}H_{10}ClNO$ o-Chlorobenzophenone oxime, α-, D 1.61 (310); β-, D 1.61 (310). m-Chlorobenzophenone oxime, α-, B or CT 1.50 (310); β-, B or CT 1.61 (310). p-Chlorobenzophenone oxime, α-, D 2.320 (310); β-, D 2.381 (310). Salicylidene-p-chloroaniline, 25° B 2.27, D 2.49 (847).

$C_{13}H_{10}Cl_2$ p-Chlorophenyl-phenylmethyl chloride (p-chlorobenzhydril chloride), B 25° 1.8₉ (327). Diphenyldichloromethane, B 17° 2.39 (210).

$C_{13}H_{10}N_2$ Carbodianil, B 18° 1.89 (212). Diphenyldiazomethane, CT 0° 1.42 (316).

$C_{13}H_{10}N_2O_2$ Benzal-p-nitroaniline, B 5.00 ± .05 (722). 2,7-Nitroaminofluorene, D 25° 6.8 (821). p-Nitrobenzalaniline, B 4.15 ± .05 (722).

$C_{13}H_{10}N_2O$ Di-p-nitrophenylmethane, B 24° 4.29 (210).

$C_{13}H_{10}O$ Benzophenone, B 20-50° 2.5 (40); 373°K 2.5 (41); B 25° 2.95 ± .03 (108, 147); B 13° 2.95 (211); B 3.00 ± .02 (293); Lqd. 50-60° 3.09 (815); B, T, Hx or CD 20° 2.95 (827). Xanthene, B 28° 1.28 (463).

$C_{13}H_{10}O_2$ 4-Hydroxybenzophenone, D 32° 3.96 (856). Phenyl benzoate, B 22° 1.81₁ ± .01 (161); CT 1.8 (208); B 30° 1.92 (572).

$C_{13}H_{10}O_3$ 4,4'-Dihydroxybenzophenone, D 32° 4.49 (856). Phenyl salicylate, B 40° 3.15 (301, 304).

$C_{13}H_{10}S$ Thiobenzophenone, B 20° 3.37 (302).

$C_{13}H_{11}BrO$ p-Bromophenyl p-tolyl ether, Hx 25° 1.98 (453); 502-518°K 2.39 ± .02 (584).

$C_{13}H_{11}Br_2N_3$ 4,4'-Dibromo-N-methyldiazoaminobenzene, B 25° 2.52 ± .04 (553).

$C_{13}H_{11}Cl$ 3-α-Naphthyl-1-chloro-1-propene, Solid, B 32° 1.27 ± .02 (800); Liquid, B 32° 1.47 ± .02 (800).

$C_{13}H_{11}N$ Benzalaniline (N-benzylideneaniline), B 25° 1.57 (588); B 1.55 ± .1 (722). Salicylideneaniline, 25° B 2.40, D 2.57 (847). 4-Nitro-4'-methoxyazobenzene, B 17° 6.5 (709). Diphenylmethane, B 20 and 50° 0.37 (40); 373°K < 0.4 (41); B 12° ca. 0 (210); B 25° 0.22 to 0.26 (296); B 25° 0.33₅ (617). α-Naphthylmethyl-ethylene, B 27° ca. 0 (464).

$C_{13}H_{12}N_2$ Benzaldehydephenylhydrazone, B 19° 1.97 (275); B 20° 1.89 (280). Benzophenonehydrazone, B 16° 2.02 (463). 2,9-Diaminofluorene, B 18° 1.96 (209).

$C_{13}H_{12}N_2O$ sym-Diphenylurea (carbanilid), D 20° 4.6 (302). unsym-Diphenylurea, D 17° 2.7 ± .1 (578). p-Methoxyazobenzene, B 23° 1.29 (463).

$C_{13}H_{12}N_2S$ sym-Diphenylthiourea (thiocarbanilid), D 20° 4.85 (302); D 26° 4.9 (578).

$C_{13}H_{12}O$ p-Methyldiphenyl ether, B 20° 1.31 (216).

$C_{13}H_{12}S$ Phenyl p-tolyl sulfide, B 25° 1.75 (526).

$C_{13}H_{13}NO$ p-Homosalicylideneaniline, red, B 25° 2.91 (641); yellow, B 25° 2.95 (641).

$C_{13}H_{13}N_3$ N-Methyldiazoaminobenzene, B 25° 1.49 ± .03 (553).

$C_{13}H_{16}O_4$ Diethyl phenylmalonate, 30° 2.543 (738).

$C_{13}H_{17}N$ 1-Methyl-3-tert-butylindole, B 25° 2.03 (797).

$C_{13}H_{17}NO_2$ Ethyl N-dimethyl-p-aminocinnamate, B 19° 4.6 (708).

$C_{13}H_{20}O_8$ Pentaerythritol tetraacetate (tetraacetylpentaerythrite), B 2.6 (23); B > 0 (35); B 25° 1.9 (61); B 22° 2.18 ± .03 (128); Vap. ca. 3 (287).

$C_{13}H_{22}O_2$ Tetracarboxymethane tetraethyl ester, B > $\underline{0}$ (35).
Bornyl propionate, $\underline{1}$-, S 22° $\underline{1.84}$ ± .05 (109).

$C_{13}H_{24}O_2$ Menthyl propionate, $\underline{1}$-, B 22° $\underline{1.77}_5$ ± .015 (161).

$C_{13}H_{24}O_4$ Diethyl azelate, Lqd. 30-40° $\underline{2.36}$ (815).
Diethyl di-n-propylmalonate, B $\underline{2.15}$ (773).

$C_{13}H_{26}O_2$ Ethyl n-undecylate, Hp 25° $\underline{1.89}$ (687).

$C_{13}H_{28}O_2$ 2,2-Di-n-propoxy-n-heptane, B 25° $\underline{0.92}$ (437).

$C_{13}H_{28}O_4$ Pentaerythritol tetraethyl ether, B $\underline{1.1}$ (23).

$C_{14}H_6Cl_2O_2$ 1,8-Dichloroanthraquinone, B 22° $\underline{2.82}$ (640).
2,3-Dichloroanthraquinone, B 22° $\underline{2.52}$ (640).

$C_{14}H_7ClO_2$ 1-Chloroanthraquinone, D 30° $\underline{1.9}$ (577); B 25°· $\underline{1.8}$, D 24° $\underline{1.9}$ (630); 22° B $\underline{1.53}$, D $\underline{1.55}$ (640).
2-Chloroanthraquinone, B 22° $\underline{1.70}$ (640).

$C_{14}H_8Cl_2$ 1,8-Dichloroanthracene, D 23° $\underline{3.2}$ (577).
2,2'-Dichlorodiphenylacetylene, B 25° $\underline{1.92}$ (824).
3,3'-Dichlorodiphenylacetylene, B 25° $\underline{1.91}$ (824).

$C_{14}H_8Cl_4$ 1,5-Dichloroanthracene 9,10-dichloride, cis-, α-methylnaphthalene 25° $\underline{3.7}$ (577).
1,8-Dichloroanthracene 9,10-dichloride, trans-, D 45° $\underline{2.4}$ (577).

$C_{14}H_8N_2$ 4,4'-Dicyanobiphenyl, B 25° $\underline{1.3}_0$ (604).

$C_{14}H_8N_2O$ Benzoylenebenzimidazole, B 25° $\underline{1.97}$ (821).

$C_{14}H_8N_6$ Biphenyl-4,4'-bisdiazocyanide, cis-, B 25° $\underline{2.8}_8$ (540); trans-, B 25° $\underline{1.8}_0$ (540).

$C_{14}H_8O_2$ Anthraquinone, B $\underline{0.6}$ (598); B 22° $\underline{0}$ (640).
Phenanthrenequinone (phenanthraquinone), S $\underline{5.6}$ (620); 25° B $\underline{5.5}_7$, D $\underline{5.66}$, CD $\underline{5.5}_0$, Cf $\underline{4.7}_1$ (634, 635).

$C_{14}H_8O_3$ Diphenic anhydride, B 25° $\underline{5.29}$ (601).

$C_{14}H_9Br$ 9-Bromoanthracene, B 25° $\underline{1.5}_0$ (127).

$C_{14}H_9Br_2Cl_3$ 2,2-Bis(p-bromophenyl)-1,1,1-trichloroethane, B 20° $\underline{1.19}$ (846).

$C_{14}H_9ClN_2O$ o-Chlorophenyl N-ether of oximinophenylacetonitrile, $\underline{\alpha}$-, B 25° $\underline{6.9}_3$ (628); $\underline{\beta}$-, B 25° $\underline{1.2}_2$ (628).
m-Chlorophenyl N-ether of oximinophenylacetonitrile, $\underline{\alpha}$-, B 25° $\underline{6.2}_2$ (628); $\underline{\beta}$-, B 25° $\underline{1.7}_8$ (628).
p-Chlorophenyl N-ether of oximinophenylacetonitrile, $\underline{\alpha}$-, B 25° $\underline{5.6}_3$ (628); $\underline{\beta}$-, B $\underline{1.5}_3$ (628).

$C_{14}H_9Cl_5$ 2,2-Bis(p-chlorophenyl)1,1,1-trichloroethane ("DDT"), B 20° $\underline{0.91}$ (833b); B 20° $\underline{1.12}$ (846).
2-p-Chlorophenyl-2-o'-chlorophenyl-1,1,1-trichloroethane, B 20° $\underline{2.24}$ (846).
2-p-Chlorophenyl-2,m'-chlorophenyl-1,1,1-trichloroethane, B 20° $\underline{1.77}$ (846).

$C_{14}H_{10}$ Anthracene, B 25° ca. $\underline{0}$ (582); B 25-30° $\underline{0}$, D 20-30° $\underline{0}$ (855).
Diphenylacetylene (tolan), B 10-70° $\underline{1.12}$ (181); B 25° $\underline{0.3}_0$ (327); B 18° $\underline{0}$ (217, 461).
Phenanthrene, Hp 20° $\underline{0}$ (220).

$C_{14}H_{10}BrCl$ 2-p-Chlorophenyl-2-phenyl-1-bromoethylene (p-chlorodiphenylvinyl bromide), low melting, B 18° $\underline{1.28}$ (211); high melting, B 17° $\underline{2.27}$ (211).

$C_{14}H_{10}Br_2$ 2-p-Bromophenyl-2-phenyl-1-bromoethylene (p-bromodiphenylvinyl bromide), low melting, B 20° $\underline{1.22}$ (211); high melting, B 21° $\underline{2.43}$ (211).
1,1-Diphenyl-2,2-dibromoethylene (α,β-dibromostilbene), B 15° $\underline{1.62}$ (211); B 20° $\underline{2.53}$ (461).

$C_{14}H_{10}Cl_2$ 1,1-Bis(p-chlorophenyl)ethylene, B 13° $\underline{1.39}$ (211); B 25° $\underline{1.43}$ (753).

1,1-Diphenyl-2,2-dichloroethylene (α, β-dichlorostilbene), B 13° $\underline{1.78}_5$(211); MP 60°, B 20° $\underline{2.69}$ (461); MP 144°, B 15° $\underline{0}$ (461).

$C_{14}H_{10}Cl_2O_2$ 1,5-Dichloro-9,10-dihydroxy-9,10-dihydroanthracene, cis-, D 25° $\underline{2.95}$ (577).

$C_{14}H_{10}F_2$ 1,1-Bis(p-fluorophenyl)ethylene, B 25° $\underline{1.52}$ (753).

$C_{14}H_{10}N_2O$ Diphenylazoxime, S $\underline{1.563}$ (433).
Diphenylfurazan (3,4-diphenyl-1,2,5-oxadiazole), S $\underline{4.739}$ (433).
2,5-Diphenyl-1,3,4-oxadiazole, S $\underline{3.86}$ (433); B 25° $\underline{3.45}$ (532).
Phenyl N-ether of oximinophenylacetonitrile, α-, B 25° $\underline{6.3}$ ± .1 (337); B 25° $\underline{6.30}$ (628); β-, B 25° $\underline{1.0}_7$ (337); B 25° $\underline{1.0}_1$ (628).

$C_{14}H_{10}N_2O_2$ Azodibenzoyl, B 25° $\underline{2.85}$ (532).
Diphenylfurazan peroxide, B 25° $\underline{5.19}$ (666).

$C_{14}H_{10}N_2O_4$ α, β-Dinitrostilbene, B 16° $\underline{0}$ (461).
p,p'-Dinitrostilbene, α-methylnaphthalene 16° ca. $\underline{0}$ (461).
1,1-Diphenyl-2,2-dinitroethylene, B 15° $\underline{5.49}$ (211).

$C_{14}H_{10}N_2O_5$ o,p'-Dinitrostilbene oxide, B 15° $\underline{4.96}$ (213).
p,p'-Dinitrostilbene oxide, low melting, D 17° $\underline{5.75}$ (213); high melting, α-methylnaphthalene 16° ca. $\underline{2.1}$ (213).

$C_{14}H_{10}N_2S_2$ 2,3-Diphenyl-2,5-endothio-1,3,4-thiadiazoline, 25° B $\underline{8.8}$, D $\underline{9.1}$ (787).

$C_{14}H_{10}O$ 1-Anthrol, B 20° $\underline{1.44}$ (549).
2-Anthrol, B 20° $\underline{1.53}$ (549).
Anthrone, B 20° $\underline{3.66}$ (583).

$C_{14}H_{10}O_2$ Benzil, B 18° $\underline{3.71}$ (78); B 25° $\underline{3.2}$ (88); B 25° $\underline{3.62}$, 50° $\underline{3.62}$, Hx 25° $\underline{3.45}$, CT 25° $\underline{3.52}$, 50° $\underline{3.46}$ (593); 25° B $\underline{3.7}_6$, Hx $\underline{3.4}_9$, Dec $\underline{3.3}_2$, D $\underline{3.7}_8$, CT $\underline{3.6}_0$, CD $\underline{3.44}$, Cf $\underline{3.2}$ (634, 635).

$C_{14}H_{10}O_2S_2$ Benzoyl persulfide, B 25° $\underline{1.1}$, 45° $\underline{1.4}$ (770).
$C_{14}H_{10}O_3$ Benzoic anhydride, B 25° $\underline{4.15}$ (770).
$C_{14}H_{10}O_4$ Benzoyl peroxide, B 45° $\underline{1.58}$ (770).
$C_{14}H_{11}Br$ 1,2-Diphenyl-1-bromoethylene (α-bromostilbene), MP < R.T., B 23° $\underline{1.30}$ (461); MP 31°, B 21° $\underline{1.38}$ (461).
2,2-Diphenyl-1-bromoethylene (diphenylvinyl bromide), B 22° $\underline{1.51}$ (211).
1-Phenyl-1-p-bromophenylethylene, B 25° $\underline{1.50}$ (753).

$C_{14}H_{11}Cl$ 1-Phenyl-1-p-chlorophenylethylene, B 25° $\underline{1.49}$ (753).
$C_{14}H_{11}Cl_3$ 2,2-Diphenyl-1,1,1-trichloroethane, B 20° $\underline{1.94}$ (846).
$C_{14}H_{11}N$ 2-Phenylindole, B 25° $\underline{2.01}$ (797).
3-Phenylindole, B 25° $\underline{2.21}$ (797).

$C_{14}H_{11}NO$ N-Methylacridone, B 30° $\underline{3.5}$ (709).
$C_{14}H_{11}NO_3$ p-Nitrostilbene oxide, B 17° $\underline{4.13}$ (213).
$C_{14}H_{11}NS$ N-Methylthioacridone, α-methylnaphthalene 17° $\underline{5.2}$ (709).
$C_{14}H_{11}N_3O_2$ 1,4-Diphenyl-3,5-dioxo-1,2,4-triazolidine, 25° B $\underline{1.98}$, D $\underline{2.71}$ (787).

$C_{14}H_{11}N_3O_4$ α-Benzoyl-β-p-nitrobenzoylhydrazide, B 25° $\underline{5.57}$ (532).
$C_{14}H_{12}$ 9,10-Dihydroanthracene, B 25° ca. $\underline{0.4}$ (582).
1,1-Diphenylethylene, B 10-70° $\underline{0.5}$ (181); B 25° $\underline{0}$ to $\underline{0.38}$ (753).
1,2-Diphenylethylene, B $\underline{0.5}$ (42); cis-, (isostilbene), B 18° $\underline{0}$ (461); trans-, (stilbene), B 18° $\underline{0}$ (111); B 10-70° $\underline{0}$ (181).

$C_{14}H_{12}Br_2$ Stilbene dibromide, dl-, B 25° $\underline{2.8}_1$ (822); meso-, B 25° $\underline{0.4}$ to $\underline{0.9}$ (822).

$C_{14}H_{12}ClN$ p-Chlorobenzylidene-p-toluidine, B 25° $\underline{2.06}$ (588).
o-Methoxybenzylidene-p-chloroaniline, B 25° $\underline{3.85}$ (847).

$C_{14}H_{12}Cl_2$ 4,4'-Dichloro-2,2'-dimethylbiphenyl, B 25° $\underline{0.75}$ (601).
Stilbene dichloride, α-, B 25° $\underline{1.2}_7$ (141, 142); 25° B $\underline{1.45}$, CT $\underline{1.32}$ (593); β-, B 25° $\underline{2.7}_5$ (141, 142).

$C_{14}H_{12}N_2$ Dibenzalhydrazine (benzylideneazine), S 0.89 (117); B 18° 1.00 (275).

$C_{14}H_{12}N_2O$ Benzoylformaldehyde-phenylhydrazone, α-, or cis-, B 20° 1.70 (595); β- or trans-, B 20° 2.72 (595).

$C_{14}H_{12}N_2O_2$ α,β-Dibenzoylhydrazide, D 25° 2.63 (532).
Diphenylglyoxime, α-, D 20° 1.492 (366); β-, D 20° 2.12 (366); γ-, D 20° 1.55 (366).

$C_{14}H_{12}N_2O_3$ Methyl N-ether of p-nitrobenzophenoneoxime, α-, B 25° 6.60 ± .1 (191); β-, B 25° 1.09 ± .1 (191).
Methyl O-ether of p-nitrobenzophenoneoxime, α-, S 25° 3.75 (323); β-, S 25° 4.26 (323).

$C_{14}H_{12}N_2O_4$ 4,4'-Dinitro-2,2'-dimethylbiphenyl, B 25° 1.30 (601).

$C_{14}H_{12}O$ Stilbene oxide, B 14° 1.73 (213).

$C_{14}H_{12}O_2$ Benzoin, B 18° 3.46 (78); B 18° 3.57, 40° 3.54, 60° 3.58 (116, 120).
Benzyl benzoate, 25° Dec 2.08, cymene 1.89 (159).
2,6-Dimethyldi-o-phenylene dioxide, B 20° 0.61 (724, 725).
2-Methyl-2-phenyl-o-phenylene methylene dioxide, S 1.05 (466).

$C_{14}H_{13}NO$ 4-Acetamidobiphenyl, B 25° 3.83 (492).
o-Methoxybenzylideneaniline, B 25° 3.02 (588).
Salicylidene-m-toluidine, B 25° 2.59 (588).
Salicylidene-p-toluidine, 25° B 2.63, D 2.74 (847).

$C_{14}H_{14}$ 2,2'-Dimethylbiphenyl, B 25° 0.66 (601).
1,2-Diphenylethane (bibenzyl), Lqd. 58-178° 0 (44); B 18° 0.36 (111); B 25° 0.45$_5$ (617).

$C_{14}H_{14}Br_2S$ Benzyl sulfide dibromide, B 25° ca. 5.4 (784).

$C_{14}H_{14}I_2S$ Benzyl sulfide diiodide, B 10° 4.4 (784).

$C_{14}H_{14}NO_3$ p,p'-Dianisyl nitroxide, S 3.3 ± .3 (277).

$C_{14}H_{14}N_2$ p-Azotoluene, trans-, B 25° ca. 0 (645).
9,10-Dimethyl-9,10-dihydrophenazine, B 25° ca. 0.4 (582).

$C_{14}H_{14}N_2O$ o,o'-Azoxytoluene, cis-, B 4.36$_5$ (225, 294); trans-, B 1.73 (225, 294).
p,p'-Azoxytoluene, cis-, B 5.06 (294); trans-, B 1.73$_5$ (294).
α-Benzoyl-β-p-tolylhydrazide, B 25° 3.38 (532).

$C_{14}H_{14}N_2O_2$ Dibenzyl hyponitrite, B 20° 0.4 (303).

$C_{14}H_{14}N_2O_3$ o,o'-Azoxyanisole, cis-, S 6.16 (225); B 6.17 (294); trans-, S 2.40 (225); B 2.41$_5$ (294).
p,p'-Azoxyanisole, Solid and Lqd. 110-202° 2.3 (37).

$C_{14}H_{14}N_4O_2$ 4-Nitro-4'-dimethylaminoazobenzene, D 18° 8.1 (709).

$C_{14}H_{14}O$ Benzyl ether, B 21° 1.38 (103).
2,2'-Dimethylphenyl ether, B 27° 0.83, 50° 0.84, Hx 31° 0.82 (647, 672).
3,3'-Dimethylphenyl ether, B 30° 1.40 (647, 671).
3,4-Dimethylphenyl ether, B 30° 1.53 (647, 671).
3,4'-Dimethylphenyl ether, B 30° 1.42 (647, 671).
4,4'-Dimethylphenyl ether (p-tolyl ether), B 25° 1.41 to 1.43 (296); 502°K 1.44 ± .03 (584); B 30° 1.46 (647, 671).

$C_{14}H_{14}OS$ Benzyl sulfoxide, B 23° 3.88 (103).
p-Tolyl sulfoxide, B 25° 4.40 (296).

$C_{14}H_{14}O_2$ o,o'-Dimethoxybiphenyl, B 25° 1.52 ± .002 (68).
p,p'-Dimethoxybiphenyl, B 25° 1.52 (34).
Hydrobenzoin (meso-), B 60° 2.33 (111); B 18° 2.33 (116); B 18° 2.33, 60° 2.31 (169); B 25° 2.0$_6$ (199); B 25° 2.48 (530); isohydrobenzoin, dl-, B 60° 2.39 (111); B 18° 2.99 (116); B 18° 2.70, 60° 2.65 (169); B 25° 2.6$_7$ (199); B 25° 2.72 (530); l-, B 25° 2.70 (530).

$C_{14}H_{14}S$ — Benzyl sulfide, B 21° 1.38 (103).
p-Tolyl sulfide, B 25° 1.92 to 1.94 (296).

$C_{14}H_{14}Se$ — p-Tolyl selenide, B 25° 1.81 (860).

$C_{14}H_{14}Se_2$ — Benzyl diselenide, B 25° 1.54 (860).
p-Tolyl diselenide, B 25° 2.29 (860).

$C_{14}H_{15}N_3$ — p-Dimethylaminoazobenzene, B 27° 3.68 (463).
4,4'-Dimethyldiazoaminobenzene, B 25° 0.9 ± .04 (553).

$C_{14}H_{16}N_2$ — 6,6'-Diamino-2,2'-dimethylbiphenyl, B 20° 1.66 (155).

$C_{14}H_{17}NO$ — p-Dimethylaminocinnamylideneacetone, MP 120-2°, B 25° 6.7 (708); MP 215°, D 20° 2.4 (708).

$C_{14}H_{18}O_4$ — Diethyl phenyl-methylmalonate, S 30° 2.52 (738).

$C_{14}H_{22}O_2$ — 2,5-Di-tert-butylhydroquinone, B 25° 1.68 ± .20 (771).
Hydroquinone di-n-butyl ether, B 25° 1.79 (94).

$C_{14}H_{24}O_2$ — Bornyl n-butyrate, l-, B 22° 1.784 ± .015 (161).

$C_{14}H_{26}O_2$ — Menthyl n-butyrate, l-, B 22° 1.664 ± .015 (161).

$C_{14}H_{26}O_4$ — Diethyl sebacate, B 25° 2.49, 50° 2.50 (185); Lqd. 30-40° 2.42 (815).

$C_{14}H_{28}O_2$ — Ethyl laurate, Lqd. 20-143° 1.3 (44).
Myristic acid, B 25° 0.76 (317); B 25° 0.77 (698).

$C_{15}H_{10}$ — Phenanthridine, B 25° 1.50 (641).

$C_{15}H_{10}Br_2O$ — α,β-Dibromobenzylideneacetophenone, B 22° 3.17 (461).
p,p'-Dibromobenzylideneacetophenone, D 17° 2.03 (461).

$C_{15}H_{10}O_2$ — 3-Phenylcoumarin, S 25° 4.30 (552).

$C_{15}H_{11}BrO$ — Benzylidene-p-bromoacetophenone, B 18° 2.93 (461).
α-Bromobenzylideneacetophenone, B 20° 3.87 (461).
β-Bromobenzylideneacetophenone, B 21° 3.59 (461).
p-Bromobenzylideneacetophenone, B 22° 2.47 (461).

$C_{15}H_{11}ClO_5$ — 2-Phenylbenzopyrylium perchlorate, dimethylaniline 25° ca. 8 (490).

$C_{15}H_{11}NO$ — α,γ-Diphenylisoxazole, B 3.33 (704).

$C_{15}H_{12}N_2O$ — o-Tolyl N-ether of oximinophenylacetonitrile, α- or cis-, B 25° 6.42 (628); β- or trans-, B 25° 0.96 (628).
m-Tolyl N-ether of oximinophenylacetonitrile, α- or cis-, B 25° 6.83 (628); β- or trans-, B 25° 1.06 (628).
p-Tolyl N-ether of oximinophenylacetonitrile, α- or cis-, B 25° 6.84 (628); β- or trans-, B 25° 1.05 (628).

$C_{15}H_{12}O$ — Benzylideneacetophenone (benzalacetophenone), B 20° 2.92 (461); MP 49°, B 20° 3.02 (531); MP 57°, B 20° 2.98 (531); MP 59°, B 20° 3.00 (531).

$C_{15}H_{12}O_2$ — Benzalacetophenone oxide, B 15° 3.86 (213).

$C_{15}H_{13}IN_2S_2$ — 5-Methylmercapto-2,3-diphenyl-1,3,4-thiadiazolinium iodide, Cf 25° 13.1 (787).

$C_{15}H_{13}NO_3$ — 4-Nitro-4'-methoxystilbene, B 17° 7.8 (709).

$C_{15}H_{13}N_3OS$ — 1,4-Diphenyl-5-methylmercapto-3,5-endoxy-1,2,4-triazoline, D 25° 7.7 (787).

$C_{15}H_{13}N_3S$ — 1,4-Diphenyl-5-methyl-3,5-endothio-1,2,4-triazoline, Cf 25° 8.4 (787).

$C_{15}H_{14}N_2$ — p,p'-Dimethylcarbodianil, B 17° 1.96 (212).
Di-p-tolyldiazomethane, CT 0° 1.96 (316).

$C_{15}H_{14}N_2S$ — 3-Ethyl-2-(phenylimino)-benzothiazoline, B 25° 1.6 ± .6 (714).

$C_{15}H_{14}O$ — Dibenzyl ketone, B 18° 2.65 (78).

$C_{15}H_{14}O_2$ — Ethyl p-phenylbenzoate, B 15° 2.01 (461).

$C_{15}H_{14}O_2S$ — Di-p-anisyl thioketone, B 25° 4.44 ± .02 (108); B 25° 4.44 ± .05 (147).

$C_{15}H_{14}O_3$ — Di-p-anisyl ketone, B 25° 3.90 ± .02 (108, 147).

$C_{15}H_{14}S$ — Dimethylsulfonium-9-fluorenylidide, B 25° 6.2 ± .2 (817).

$C_{15}H_{15}N_3O_2$ — p-Dimethylaminobenzal-p'-nitroaniline, B 8.6 ± .2 (722).

p-Nitrobenzal-p'-dimethylaminoaniline, B 6.87 ± .05 (722).

$C_{15}H_{16}$ 1,3-Diphenylpropane, B 25° 0.55 (617).

$C_{15}H_{16}N_2$ Benzal-p-dimethylaminoaniline, B 2.65 ± .1 (722).

p-Dimethylaminobenzalaniline, B 3.6 ± .05 (722).

sym-Diphenyl-dimethylurea, D 24° 3.6 ± .1 (578).

$C_{15}H_{16}O_2$ Di-p-anisylmethane, B 25° 1.61 ± .07 (108, 147).

$C_{15}H_{20}Br_2$ Dihydro-α-tricyclopentadiene-1,2-dibromide, cis-, B 3.20 (223).

Dihydro-β-tricyclopentadiene-1,2-dibromide, cis-, B 3.18 (223); trans-, B 1.92 (223).

$C_{15}H_{20}O_4$ Diethyl β-phenylglutarate, S 30° 2.505 (738).

$C_{15}H_{22}O$ Aromadendrone, B 23° 2.11 (696).

Vetivone, α- or cis-, S 23° 3.87 (736); β- or trans-, S 23° 3.73 (736).

$C_{15}H_{22}O_3$ Bithone, ethyl alcohol 22° 3.71 (571).

$C_{15}H_{22}O_{10}$ Tetraacetyl-methylglucoside, α-, S 2.38 (117); β-, S 3.08 (117).

$C_{15}H_{24}$ Aromadendrene, B 23° 0.93 (696).

Cedrene, Lqd. 15-35° 0.38 (649).

Sesquiterpene from α- or cis-vetivone, S 23° 1.12 (736); from β- or trans-vetivone, S 23° 0.97 (736).

$C_{15}H_{26}$ Dihydroaromadendrene, B 23° 0.79 (696).

$C_{15}H_{28}$ Vetivane, α- or cis-, S 23° 0.93 (736); β- or trans-, S 23° 0.73 (736).

$C_{15}H_{30}O_2$ Methyl myristate, B 25° 1.61 (317).

$C_{16}H_{12}Br_2O_4$ Dimethyl 4,4'-dibromodiphenate, B 25° 2.17 (601).

$C_{16}H_{12}N_2O$ 1-Benzeneazo-2-naphthol, CD 16° 1.4 (463); B 25° 1.60 (645).

2-Benzeneazo-1-naphthol, B 28° 1.8 (463).

4-Benzeneazo-1-naphthol, D 28° 2.1 (463).

$C_{16}H_{12}N_2O_8$ Dimethyl 4,4'-dinitrodiphenate, B 25° 2.03 (601).

$C_{16}H_{13}ClO_5$ 2-Phenyl-3-methylbenzopyrylium perchlorate, dimethylaniline 25° ca. 7 (490).

$C_{16}H_{13}NO$ 2-Methyl-4,5-diphenyloxazole, B 25° 1.7 (768).

$C_{16}H_{13}NO_4$ Methyl p-nitro-α-phenylcinnamate, B 17° 3.78 (461).

$C_{16}H_{13}N_3$ 1-Benzeneazo-2-naphthylamine, B 16° 2.14 (463).

4-Benzeneaso-1-naphthylamine, B 16° 2.56 (463).

$C_{16}H_{14}O_2$ 4,4'-Diacetylbiphenyl, B 25° 1.9 (61).

Methyl α-phenylcinnamate, B 17° 3.78 (461).

$C_{16}H_{14}O_4$ 2,2'-Dihydroxybiphenyl diacetate (2,2'-biphenol diacetate), B 25° 2.18 (68).

4,4'-Dihydroxybiphenyl diacetate (4,4'-biphenol diacetate), B 25° 1.9 (95).

Dimethyl 2,2'-biphenyldicarboxylate (dimethyl diphenate), B 25° 2.36 (68); B 25° 2.3 (95); B 25° 2.42 (601).

Dimethyl 4,4'-biphenyldicarboxylate, B 25° 2.2 (95).

$C_{16}H_{15}Cl_3$ 2,2-Di-p-tolyl-1,1,1-trichloroethane, B 20° 2.19 (846).

$C_{16}H_{15}Cl_3O_2$ 2,2-Di-p-anisyl-1,1,1-trichloroethane, B 20° 2.87 (846).

$C_{16}H_{15}N$ 1,2-Dimethyl-3-phenylindole, B 25° 2.61 (797).

1,3-Dimethyl-2-phenylindole, B 25° 2.18 (797).

2-Methyl-3-benzylindole, B 25° 2.49 (797).

3-Methyl-2-benzylindole, B 25° 2.11 (797).

$C_{16}H_{15}NO_3$ Ethyl salicylidene-p-aminobenzoate, red, B 25° 2.69 (641); B 25° 2.72 (729); yellow, B 25° 2.69 (641); B 25° 2.72 (729).

$C_{16}H_{16}$ 1,1-Diphenyl-2,2-dimethylethylene, 30° B 0.51, CT 0.65 (590).

$C_{16}H_{16}N_2O_2$ 4-Nitro-4'-dimethylaminostilbene, α-methylnaphthalene 17° 8.3 (709).

$C_{16}H_{16}N_3S_2I$ 3-Methylmercapto-5-methyl-1,4-diphenyl-1,2,4-triazolinium iodide, Cf 25° 11.0 (787).

$C_{16}H_{16}O_2$ Ethyl diphenylacetate, S 30° 1.76 (738).

$C_{16}H_{18}$ 1,4-Diphenylbutane, B 25° 0.52 (617).

$C_{16}H_{18}O_2$ 4,4'-Diethoxybiphenyl (4,4'-biphenyl diethyl ether), B 25° 1.9 (61, 65, 95).

$C_{16}H_{20}N_2$ N-Tetramethylbenzidine, B 25° 1.25 (93).

$C_{16}H_{22}O_{11}$ d-Glucose pentaacetate (pentaacetyl-d-glucose), 20° B 2.43, Cf 2.23 (383); α-, S 3.47 (117); β-, B 20° 2.405, 35° 2.469, 50° 2.518, Cf 20° 2.256, 50° 2.294 (597).

$C_{16}H_{26}O_2$ 2,5-Di-tert-butylhydroquinone dimethyl ether, B 25° 1.47 ± .1? (771).

$C_{16}H_{32}O_2$ Palmitic acid, B 25° 0.72 (317); B 25° 0.77 (698); D 25° 1.75 (752); 23° B 0.76, CH 0.81, D 1.75 (819).

$C_{16}H_{32}O_5$ Aleuritic acid (8,9,15-trihydroxypalmitic acid), D 40° 4.28 (752).

$C_{16}H_{33}I$ 1-Iodohexadecane (cetyl iodide), CT 20° 1.81 (750).

$C_{16}H_{34}O$ Cetyl alcohol, B 25° 1.6$_8$ (127); 20-60° B 1.69 to 1.70, Hx 1.40 to 1.43, E 0-20° 1.79 (300); E 20° 1.81 (358).

$C_{16}H_{36}BrN$ Tetra-n-butylammonium bromide, B 25° 11.6 (472).

$C_{16}H_{36}ClNO_4$ Tetra-n-butylammonium perchlorate, B 25° 14.1 (472).

$C_{17}H_{10}O$ Benzanthrone, B 25° 3.49 (661).

$C_{17}H_{12}OS$ 2,6-Diphenylthio-γ-pyrone, S 4.41 (374); B 20° 4.39 (402).

$C_{17}H_{12}O_2$ 2,6-Diphenyl-γ-pyrone, B 20° 3.82 (302).

$C_{17}H_{12}O_3S$ 2,6-Diphenylthio-γ-pyrone 1-dioxide, B 20° 0.93 (402).

$C_{17}H_{14}N_2O$ 4-Benzeneazo-1-methoxynaphthalene, B 25° 0.93 (463). 1-p-Tolueneazo-β-naphthol, B 25° 1.66 (645).

$C_{17}H_{14}O$ Dibenzalacetone, B 3.28 (119).

$C_{17}H_{16}N_2S$ 3-Ethyl-2-(phenyliminoethylidene)-benzothiazoline, B 25° 2.0 ± .6 (714).

$C_{17}H_{16}OS$ 2,6-Diphenylthio-γ-pyran-4-one, cis-, B 20° 1.64 (402); trans-, B 20° 1.62 (402).

$C_{17}H_{16}O_2$ β-Ethoxybenzalacetophenone, MP 63°, B 20° 3.30 (531); MP 74°, B 20° 3.28 (531); MP 78°, B 20° 3.27 (531); MP 81°, B 20° 3.28 (531). Ethyl β-phenylcinnamate, B 19° 1.98 (461).

$C_{17}H_{17}BrN_2O_2$ 4-Bromo-3'-nitro-4'-piperidino-biphenyl, B 25° 3.30 (492).

$C_{17}H_{18}N_2OS_2$ 5-Ethoxy-5-methylmercapto-2,3-diphenyl-1,3,4-thiadiazoline, B 25° 1.83 (787).

$C_{17}H_{18}O_2$ Thymol phenyl ketone, 32° B 3.34, D 3.59 (856).

$C_{17}H_{20}$ 1,5-Diphenylpentane, B 25° 0.48 (617).

$C_{17}H_{25}N_3O$ 6-Methoxy-8-(3-diethylamino)propyl-aminoquinoline (plasmocid), B 1.61 (742).

$C_{17}H_{34}O$ Di-n-octyl ketone, Lqd. 51° 2.38 (558).

$C_{17}H_{34}O_4$ Monomyristin, B 30° 2.99 (780).

$C_{17}H_{34}O_5$ Methyl aleurate (methyl 8,9,15-trihydroxypalmitate), D 25° 4.27 (752).

$C_{18}H_{10}Cl_2$ 6,12-Dichloronaphthacene, B 20° 2.5 (855).

$C_{18}H_{10}O_2$ Naphthacene-6,11-quinone, B 25° 2.3 (855).

$C_{18}H_{11}NO_2$ Quinolein yellow, D 14° 3.64 (463).

$C_{18}H_{12}$ Naphthacene, B 27° 0 (855).

$C_{18}H_{12}Br_2O_2$ Hydroquinone di-p-bromophenyl ether, B 10° 0.92, 25° 0.89, 40° 0.88 (683).

$C_{18}H_{12}Cl_3OP$ Tri-p-chlorophenylphosphine oxide, B 25° 2.93 ± .03 (817).

$C_{18}H_{12}Cl_3P$ Tri-p-chlorophenylphosphine, B 25° 0.65 ± .07 (817).

$C_{18}H_{12}N_5O_6$ α,α-Diphenyl-β-picrylhydrazyl, B 25° 4.92 (775).

$C_{18}H_{13}N_5O_6$ α,α-Diphenyl-β-picrylhydrazine, B 25° 3.59 (775).

$C_{18}H_{14}O_2$ Hydroquinone diphenyl ether, B 10-40° 1.42 (683).

$C_{18}H_{15}N$ Triphenylamine, B 15° 0.26 (214).

$C_{18}H_{15}OP$ — Triphenylphosphine oxide, B 25° 4.31 (786); B 25° 4.28 ± .02 (817).

$C_{18}H_{15}O_3P$ — Triphenyl phosphite, B 25° 2.02 (687).

$C_{18}H_{15}O_3PS$ — Triphenyl thiophosphate, B 25° 2.58 (687).

$C_{18}H_{15}O_4P$ — Triphenyl phosphate, B 20° 2.79 (412); B 25° 2.81 (687).

$C_{18}H_{15}P$ — Triphenylphosphine, B 17° 1.45 (214); B 25° 1.39 ± .04 (817).

$C_{18}H_{15}PS$ — Triphenylphosphine sulfide, B 25° 4.74 (786); B 25° 4.73 ± .02 (817).

$C_{18}H_{15}PSe$ — Triphenylphosphine selenide, B 25° 4.83 (786).

$C_{18}H_{16}O_4$ — Dimethyl 9,10-dihydroanthracene-9,10-dicarboxylate, cis-, B 24° 2.6 (577); trans-, B 24° 1.7 (577).

$C_{18}H_{17}O_2P$ — Triphenylphosphine oxide hydrate, B 25° 4.56 (786).

$C_{18}H_{18}O_4$ — 4,4'-Diethoxybenzil, B 25° 5.4 (88).

$C_{18}H_{19}Cl_3$ — 2,2-Di-2,4-dimethylphenyl-1,1,1-trichloroethane, B 20° 2.50 (846).

2,2-Di-2,5-dimethylphenyl-1,1,1-trichloroethane, B 20° 1.94 (846).

2,2-Di-3,4-dimethylphenyl-1,1,1-trichloroethane, B 20° 2.49 (846).

2,2-Di-p-ethylphenyl-1,1,1-trichloroethane, B 20° 2.19 (846).

$C_{18}H_{19}Cl_3O_2$ — 2,2-Di-p-phenetyl-1,1,1-trichloroethane, B 20° 2.97 (846).

$C_{18}H_{22}$ — 1,6-Diphenylhexane, B 25° 0.52 (617).

$C_{18}H_{27}N_3$ — 8-(1-Methyl-4-diethylamino)butylaminoquinoline (neoplasmo-quine), B 0.70 (742).

$C_{18}H_{30}N_4O_7$ — Tri-n-butylammonium picrate, B 25° 13.1 (472).

$C_{18}H_{32}O_2$ — Linoleic acid, B 30° 1.208 (614); B 1.512, T 1.340 (705); B 2° 1.512 (706); 18° Hx 1.39, CH 1.35, me. CH 1.204 (804); 23° D 1.71, CH 1.35 (805, 819); D 19° 1.71 (808).

$C_{18}H_{34}O_2$ — Oleic acid, B 30° 1.009 (614); B 1.425 (705); B 2° 1.452 (706); Hx 18° 1.22, 40° 1.16, CH 18° 1.13, me. CH 18° 1.09 (804); 23° D 1.68, CH 1.13 (805, 819); D 19° 1.68 (808); Lqd. 0.76, after 8 hrs. elect. discharge in H_2 1.08 (802); CH 25° 1.12, 50° 1.34, 70° 1.51, 79° 1.60 (862).

$C_{18}H_{36}O$ — Oleyl alcohol (cis), B 20° 1.72 (267); elaidic alcohol (trans), B 20° 1.70 (267).

$C_{18}H_{36}O_2$ — Ethyl palmitate, Lqd. 30-182° 1.2 (44); Hp 25° 1.87 (687). Stearic acid, D 25° 1.74 (399); D 2° 1.506 (706); 23° D 1.66, CH 1.04 (805, 819); CH 25° 0.67, 50° 0.78, 70° 0.89, 79° 1.05 (862).

$C_{18}H_{36}O_5$ — Ethyl aleurate, D 25° 4.31 (752).

$C_{18}H_{39}NO_2$ — Tetra-n-butylammonium acetate, B 25° 11.2 (472).

$C_{19}H_{13}N_3O_6$ — Tri-p-nitrophenylmethane, D 16° 3.23 (210).

$C_{19}H_{14}ClN$ — Diphenylmethylene-p-chloroaniline (benzophenone-p-chloranil), B 12° 2.91 (211).

$C_{19}H_{14}Cl_2$ — p-Chlorotriphenylchloromethane, B 17° 1.92_5 (210).

$C_{19}H_{14}N_2$ — 9-Fluorenonephenylhydrazone, B 16° 2.12 (463).

$C_{19}H_{14}N_2O_2$ — p-Benzoyloxyazobenzene, B 30° 1.93 (572).

$C_{19}H_{14}O$ — Fuchsone, 32° B 5.80, D 5.83 (856).

$C_{19}H_{14}O_2$ — Benzaurin, D 32° 6.85 (856).

$C_{19}H_{14}O_3$ — Aurin, D 32° 7.96 (856).

$C_{19}H_{15}$ — Triphenylmethyl ⇌ hexaphenylethane, B 8-35° < 0.7 (681).

$C_{19}H_{15}Br$ — Triphenylmethyl bromide, B 25° 2.1 (810).

$C_{19}H_{15}Cl$ — Triphenylchloromethane, B 10-70° 1.95 (180); CT 1.9 (208); B 17° 1.92_5 (210).

$C_{19}H_{15}N$ — Diphenylmethyleneaniline (benzophenoneanil), B 13° 1.97 (211); B 25° 2.0_3 (641).

$C_{19}H_{16}$ — Triphenylmethane, Lqd. 94-175° 0 (44); CD 15° 0.62 (210);

B $\underline{0.21}$, D $\underline{0.46}$ (806).

$C_{19}H_{16}N_2$ Benzophenonephenylhydrazone, B 15° $\underline{2.22}$ (463).

$C_{19}H_{16}O$ Triphenylcarbinol, B 10-70° $\underline{2.11}$ (180).

$C_{19}H_{19}NO$ p-Dimethylaminocinnamylideneacetophenone, D 26° $\underline{5.4}$ (708).

$C_{19}H_{20}N_2S$ 3-Ethyl-2-(phenyliminobutenylidene)benzothiazoline, B 25° $\underline{2.0} \pm .6$ (714).

$C_{19}H_{20}O_2$ 2,6-Diphenyl-3,5-dimethyltetrahydro-γ-pyrone, S 25° $\underline{1.80}$ (552).

$C_{19}H_{20}O_4$ Diethyl diphenylmalonate, S 30° $\underline{4.433}$ (738).

$C_{19}H_{26}O_2$ Δ^4-Androstene-3,17-dione, D 25° $\underline{3.32}$ (812).

$C_{19}H_{28}O_2$ Androstane-3,17-dione, D 25° $\underline{3.25}$ (811).

Δ^5-Androstene-3 cis-ol-17-one, D 25° $\underline{2.46}$ (812).

Testerone, D 25° $\underline{4.32}$ (812); cis-, D 25° $\underline{5.17}$ (812).

$C_{19}H_{30}O_2$ Δ^5-Androstene-3 cis, 17 cis-diol, D 25° $\underline{2.69}$ (812); 3 cis, 17 trans-diol, D 25° $\underline{2.89}$ (812).

Androsterone, D 25° $\underline{3.70}$ (812); β-, D 25° $\underline{3.73}$ (812).

$C_{19}H_{32}O_2$ Androstane-3 trans, 17 trans-diol, D 25° $\underline{2.29}$ (811).

Androstane-3 cis, 17 trans-diol, D 25° $\underline{2.99}$ (811).

$C_{20}H_{12}$ Perylene, B 25° $\underline{1.3}$, 27° $\underline{1.9}$, 30° $\underline{2.1}$ (855).

$C_{20}H_{15}Br_2Cl$ α,β-Diphenyl-p-chlorostyryl dibromide, MP 113°, B 16° $\underline{1.57}$ (461); MP 160°, B 17° $\underline{2.61}$ (461).

$C_{20}H_{16}$ Triphenylethylene, B 10-70° $\underline{0.6}$ (181).

$C_{20}H_{16}Cl_2$ 1,4-Bis(α-chlorobenzyl)benzene, dl-, B 25° $\underline{2.4}_8$ (198, 264, 327); meso-, B 25° $\underline{2.2}_8$ (198, 264, 327).

α,β-Diphenylstyryl dichloride, B 18° $\underline{1.53}$ (461).

$C_{20}H_{16}N_4$ 4,5-Dihydro-1,4-diphenyl-3,5-phenylimino-1,2,4-triazole (nitron), B 30° $\underline{7.2}$ (623).

$C_{20}H_{18}$ 1,1,1-Triphenylethane, B 10-70° $\underline{0.4}$ (181).

$C_{20}H_{18}O_2$ Hydroquinone di-p-tolyl ether, B 10-40° $\underline{1.80}$ (683).

$C_{20}H_{26}$ 1,8-Diphenyloctane, B 25° $\underline{0.50}$ (617).

$C_{20}H_{30}O_2$ 17-Methyltesterone, D 25° $\underline{4.17}$ (811).

$C_{20}H_{32}O_2$ 17-Methyl-Δ^5-androstene-3 cis, 17 trans-diol, D 25° $\underline{2.78}$ (811).

$C_{20}H_{38}O_2$ Ethyl oleate, Lqd. 28-150° $\underline{1.35}$ (44).

$C_{20}H_{40}O_2$ Ethyl stearate, Lqd. 48-167° $\underline{1.2}$ (44); Lqd. 40-50° $\underline{1.65}$ (815).

$C_{20}H_{44}BrN$ Tetraisoamylammonium bromide, B 25° $\underline{14.7}$ (306); B 25° $> \underline{14}$ (359).

$C_{21}H_{18}O_2$ o-Cresolbenzein, D 32° $\underline{6.70}$ (856).

$C_{21}H_{24}O_4$ Diethyl β,β-diphenylglutarate, S 30° $\underline{2.43}$ (738).

$C_{21}H_{30}O$ $\Delta^{4,6}$-3,17-Dimethylandrostadiene-17 trans-ol, D 25° $\underline{1.81}$ (811).

$C_{21}H_{34}O$ 3,17-Dimethylandrostane-17 trans-ol, D 25° $\underline{2.14}$ (811).

$C_{21}H_{34}O_2$ 3,17-Dimethylandrostane-3 cis, 17 trans-diol, D 25° $\underline{2.39}$ (811); 3 trans, 17 trans-diol, D 25° $\underline{2.14}$ (811).

$C_{21}H_{36}N_4O_7$ Triisoamylammonium picrate, B 25° $\underline{12.91}$ (306); B 25° $\underline{13.9}$ (359); B 25° $\underline{13.3}$ (472).

$C_{21}H_{42}O_3$ 2-Methoxyethanol stearate, Lqd. 50° $\underline{2.08}$ (815).

$C_{21}H_{42}O_4$ Glycerol 1-monostearate (monostearin), B 30° $\underline{3.04}$ (780).

$C_{21}H_{44}N_2S$ Tetraisoamylammonium thiocyanate, B 25° $\underline{15.4}$ (472).

$C_{22}H_{16}O_2$ 6,12-Diacetonaphthacene, B 25° $\underline{3.3}_3$, 27° $\underline{3.3}_8$ (855).

$C_{22}H_{22}O_2$ Thymolbenzein, D 32° $\underline{6.67}$ (856).

$C_{22}H_{38}N_4O_7$ Tetra-n-butylammonium picrate, B 25° $\underline{17.8}$ (472).

$C_{22}H_{42}O_4$ Diethyl hexadecamethylenedicarboxylate, B 25° $\underline{2.49}$, 50° $\underline{2.48}$ (185).

$C_{23}H_{18}N_2O$ 1,2-Naphthoquinone-2-benzyl-phenylhydrazone, B 29° $\underline{2.2}$ (463).

$C_{23}H_{24}$ 1,3,5-Triphenylpentane, B 25° $\underline{0.98}_5$ (617).

$C_{23}H_{26}N_2$ 4,4'-Bisdimethylaminotriphenylmethane (leucomalachite

	green), B 25° $\underline{1.5}_7$ (642).
$C_{23}H_{44}O_2$	n-Tricosane-8,16-dione, S $\underline{3.6}$ (162).
$C_{23}H_{46}O$	Di-n-undecyl ketone, Lqd. 69° $\underline{2.48}$ (558).
$C_{24}H_{20}NP$	Triphenylphosphinephenylimide, B 25° $\underline{4.45}$ (786); B 25° $\underline{4.40}$ ± .03 (817).
$C_{24}H_{25}N_3$	4,4'-Bisdimethylaminotriphenylmethyl cyanide (malachite green leucocyanide), B 25° $\underline{1.1}_3$ (642).
$C_{24}H_{34}O_5$	Dehydrocholic acid, D 25° $\underline{5.63}$ (764).
$C_{24}H_{36}O_4$	Dehydrodesoxycholic acid, D 25° $\underline{4.82}$ (764).
$C_{24}H_{36}O_5$	Reductodehydrocholic acid, D 25° $\underline{5.16}$ (764).
$C_{24}H_{38}O_3$	Dehydrolithocholic acid, D 25° $\underline{3.72}$ (764).
$C_{24}H_{38}O_4$	Apocholic acid, D 25° $\underline{2.98}$ (764).
	3-Hydroxy-12-ketocholanic acid, D 25° $\underline{4.26}$ (764).
$C_{24}H_{40}O_3$	Lithocholic acid, D 25° $\underline{2.50}$ (764).
$C_{24}H_{40}O_4$	Desoxycholic acid, D 25° $\underline{3.22}$ (764).
	Hypodesoxycholic acid (hyodeoxycholic acid), D 25° $\underline{3.12}$ (764).
$C_{24}H_{40}O_5$	Cholic acid, D 25° $\underline{3.84}$ (764).
$C_{26}H_{14}F_2$	2,2'-Difluorobisdiphenylene-ethylene, mostly cis-, B 15° $\underline{2.51}$ (403).
$C_{26}H_{16}$	Bisdiphenylene-ethylene, B 16° $\underline{0}$ (403).
$C_{26}H_{18}Cl_2$	9,10-Diphenylanthracene 9,10-dichloride, cis-, B 38° $\underline{3.0}$ (577).
$C_{26}H_{20}$	Tetraphenylethylene, B 10-70° $\underline{0}$ (181).
$C_{26}H_{20}N_4$	1,4,5-Triphenyl-3,5-endoanilo-1,2,4-triazoline, 25° B $\underline{9.1}$, D $\underline{8.8}$ (787).
$C_{26}H_{42}O_4$	Gitogenin, D 25° $\underline{2.64}$ (811).
$C_{26}H_{46}N_4O_7$	Tetraisoamylammonium picrate, B 25° $\underline{18.00}$ (306); B 25° $\underline{19.4}$ (359); B 25° $\underline{18.3}$ (472).
$C_{27}H_{18}Br_2$	α,γ-Di-p-bromophenyl-α,γ-diphenylallene, B 19° $\underline{1.92}$ (404).
$C_{27}H_{18}Cl_2$	α,α-Di-p-chlorophenyl-γ,γ-diphenylallene, B 25° $\underline{1.57}$ (404).
$C_{27}H_{18}O_2$	α-Naphtholbenzein, 32° B $\underline{5.76}$, D $\underline{6.07}$ (856).
$C_{27}H_{19}Cl$	p-Chlorotetraphenylallene, B 22° $\underline{1.55}$ (404).
$C_{27}H_{20}$	Tetraphenylallene, B 19° $\underline{0}$ (404).
$C_{27}H_{20}N_2$	2,3-Diphenylindonephenylhydrazone, D 14° $\underline{1.93}$ (463).
$C_{27}H_{20}O_3$	α-Naphtholbenzein (phenyl-bis-(4-hydroxynaphthyl(1)carbinol), B 25° $\underline{4.3}$ (639).
$C_{27}H_{32}$	1,5,9-Triphenylnonane, B 25° $\underline{0.85}$ (617).
$C_{27}H_{32}O_3$	Thymolbenzein, B 25° $\underline{6.8}$ (639).
$C_{27}H_{44}O_3$	Tigogenin, D 25° 2.36 (811).
$C_{27}H_{44}O_4$	Chlorogenin, D 25° $\underline{2.67}$ (811).
$C_{27}H_{46}O$	Cholesterol, D 25° $\underline{1.99}$ (811).
$C_{27}H_{46}O_2$	Cholestane-3 cis-ol-7-one, D 25° $\underline{2.98}$ (812).
	Δ^5-Cholestene-3 cis-ol-7-one, D 25° $\underline{3.79}$ (812).
$C_{27}H_{48}O$	Dihydrocholesterol, D 25° $\underline{1.81}$ (811).
$C_{27}H_{48}O_2$	Cholestane-3 cis, 7 cis-diol, D 25° $\underline{2.55}$ (812); 3 cis, 7 trans-diol, D 25° $\underline{2.31}$ (812).
$C_{28}H_{33}N_3$	4,4'-Bis-diethylaminotriphenylmethyl cyanide (brilliant green leucocyanide), B 25° $\underline{1.7}_2$ (642).
$C_{28}H_{38}O_{19}$	Cellobiose octaacetate, α-, Cf 20° $\underline{2.75}$ (383); Cf 20° $\underline{2.767}$, 35° $\underline{2.846}$, 50° $\underline{3.083}$ (597).
$C_{29}H_{58}O$	Di-n-tetradecyl ketone (palmitone), B 25° $\underline{2.12}$ (698).
$C_{30}H_{50}O$	Friedelin, B 32° $\underline{2.80}$ ± .03 (800).
$C_{30}H_{50}O_2$	Cerin (a triterpene hydroxyketone), B 50° $\underline{2.39}$ ± .05 (800).
$C_{30}H_{52}O$	Friedelinol, low melting, B 32° $\underline{1.78}$ ± .07 (800); high melting, B 50° $\underline{1.81}$ ± .05 (800).
$C_{31}H_{32}$	1,3,5,7-Tetraphenylheptane, B 25° $\underline{1.51}$ (617).
$C_{34}H_{35}ClO_{13}S_4$	1-Chloro-2,3,4,6-tetra-p-toluenesulfonyl-d-glucoside

(1-chloro-2,3,4,6-tetratosylglucose), B 20° 6.40 (676).

$C_{37}H_{27}$ Tribiphenylmethyl, B 8-35°< 0.7 (681).

$C_{38}H_{30}$ Hexaphenylethane, see under triphenylmethyl ($C_{19}H_{15}$).

$C_{39}H_{74}O_6$ Glycerol trilaurate (trilaurin), B 30° 2.59 (698, 780).

$C_{44}H_{30}N_4$ α,β,γ,δ-Tetraphenylporphine, B 25° ca. 0 (762).

$C_{51}H_{98}O_6$ Glycerol tripalmitate (tripalmitin), 23° B 2.77, D 2.90 (819).

$C_{57}H_{104}O_6$ Glycerol trioleate (triolein), B 30° 3.5158 (614); B 2° 3.124 (705, 706); 23° B 3.08, D 3.06 (805, 819); D 19° 3.06 (808).

$C_{57}H_{104}O_9$ Glycerol triricinoleate (triricinolein), B 29° 4.117 (614).

$C_{57}H_{110}O_6$ Glycerol tristearate (tristearin), B 35° 2.7 (188); B 29° 2.844 (614); B 30° 2.70 (780); 23° B 2.83, D 2.95 (805, 819).

Oils

Castor, B 30° 3.7 (188); B 26° 3.68 (440); B 3.80 (465); gasoline 25° 3.62 (835).

Coconut, B 26° 2.82 (440); B 2.79 (465).

Linseed, B 27° 3.0 (188); B 2.99 (465).

Lubricating, mineral, Lqd. 20° 0.22 to 0.41, 50° 0.24 to 0.49, B 20° 0.22 to 0.40, 50° 0.20 to 0.45 (720).

Olive, B 26° 3.03 (440); B 3.01 (465); Lqd. 20° 2.25, 50° 2.30, B 20° 2.26 to 2.36, 50° 2.30 to 2.35 (720).

Peanut, Lqd. 20° 2.30, 50° 2.35, B 20° 2.31 to 2.39, 50° 2.35 to 2.41 (720).

Poppy, B 3.06 (465).

Rape, B 2.74 (465); Lqd. 20° 2.59, 50° 2.72, B 20° 2.61 to 2.78, 50° 2.73 to 2.85 (720).

Sesame, B 26° 2.91 (440); B 2.93 (465).

Tung, B 2.29 (157); B 25° 2.8 (188); B 20° 2.18, 60° 2.15, 100° 2.02 (478); kerosene 25° 2.92 (835).

Polymers

$(C_2H_4O)_n$ "Carbowax 4000" (linear condensation polymer of ethylene glycol), B 44° (assuming av. MW 3750) 9.91 (820).

$(C_4H_5Cl)_n$ Polychloroprene (linear addition polymer of 2-chloro-1,3-butadiene), S 1.42 (685); "Neoprene GN", solid 20-60° 1.99 per unit (818).

$(C_4H_6O_2)_n$ Polyvinyl acetate, S (MW 24,200) 28.7, (MW 60,000) 44.4 (685).

$(C_5H_8)_n$ Rubber (caoutchouc), B 25° 2.45; gel fraction, B 25° 2.91; sol fraction, 25° B 2.79, E 0.72 (423); crepe, lightly milled, B 27; more strongly milled, B 33 (250); pale crepe, 1 x milled, MW 20,000-200,000, B 14.47 to 45.74, 2 x milled, MW 20,000-200,000, B 18.38 to 58.13, 2 x milled and reprec. B 17.65 to 55.79 (436); smoked sheet, 1 x milled B 14.56 to 46.05 (436); acetone-extracted in N_2, B 25° 0.71; same plus ultraviolet irradiation, B 25° 1.10 (484).

$(C_5H_8O_2)_n$ Polymethyl methacrylate, 1.3 per unit (731).

$(C_5H_8O_2)_m-$
$\quad(C_8H_8)_n$ Methyl methacrylate-styrene copolymer (34 and 17% methyl methacrylate by wt.), B 20° 1.64 and 1.65 per methacrylate unit (760).

$(C_8H_8)_n$ Polystyrene, (MW 2820) 6.32, (MW 7750) 7.98, (MW 23,000) 8.61 (164); S (MW 12,900) 2.9, (MW 42,900) 5.7 (685).

$(C_9H_8)_n$ Polyindene, B 25° (MW 116) 0.40, (MW 855) 1.68, (MW 1490) 2.11 (164); S (MW 603) 1.37, (MW 855) 1.74, (MW 1039) 1.73, (MW 1135) 1.91, (MW 1490) 2.18 (685).

$(C_{10}H_{18}O_2)_n$ 10-Hydroxy-n-decanoic acid polymer, B 25° (n=5) 5.0, (n=12) 6.7, (n=24) 10.2, (n=46) 12.4, (n=53) 15.7, (n=82) 19.0 (523).

$(C_{12}H_{16}O_8)_n$ Triacetylcellulose (cellulose triacetate), B 20° 1.95 per glucose unit (448).

$(C_{12}H_{22}O_5)_n$ Triethylcellulose, B 20° 3.26 per glucose unit (448); B 20° 3.26 per glucose unit, 35° 3.69, 50° 3.92 (597);

MW 5,000-100,000, B 9.82 to 43.9 (436).

$(C_{27}H_{28}O_5)_n$ Tribenzylcellulose, B 20° 2.56 per glucose unit (448); B 20° 2.56 per glucose unit, 35° 2.64, 50° 2.69 (597); MW 5,000-100,000, B 9.65 to 42.8 (436).

Proteins   Egg albumin, water 25° 250 (793).

Carboxyhemoglobin, water 25° 4.70 or 5.00 (613); horse, water 25° 480 (793); pig, water 20° 410 (793).

Edestin, water 1400 (697, 793).

Gelatin (glutin), water 22° 46 (226).

Gliadin, ethyl alcohol 20° 13.5 (517); aq. eth. alc. 25° 190 (793).

Hemoglobin, water 11° 47 (226).

Insulin, 25° aq. propylene glycol 80% 360, 90% 310, 100% 300 (793).

Lactoglobulin, β -, aq. glycine 25° 730, 0° 770 (716, 793); glycine 790 ± 26 (803).

Myoglobulin, water 25° 170 (793).

Ovalbumin, water 18° 44 (226).

Secalin, aq. eth. alc. 25° 440 (793).

Serum albumin, water 380 (697); water 18° 40 (226); low polarity, water 25° 280, high polarity, water 25° 510 (586).

Serum pseudoglobulin-γ, water 1200 (697, 586); water 25° 1100, 0° 1300 (793).

Zein, aq. n-propyl alcohol 0-80° 60 (205); aq. eth. alc. 380 (793).

Miscellaneous   Gasoline, 25° Lqd. 0.17, BP 60-70° 0.15, BP 100-120° 0.22, 160-190° 0.17 (865).

Rosin, B 25° 8.8 (834).

Bibliography

(An author index to these reference numbers
may be found at the end of this bibliography.)

1919

1 M. Jona, Physik. Z. 20, 14.
2 H. Riegger, Ann. Physik (4) 59, 753.

1921

3 O. E. Frivold, Physik. Z. 22, 603.
4 P. Lertes, Z. Physik 6, 56.
5 H. Weigt, Physik.. Z. 22, 643.

1922

6 H. Falkenhagen, ibid. 23, 87.
7 J. Herweg and W. Pötzsch, Z. Physik 8, 1.
8 H. Isnardi, ibid. 9, 153.

1923

9 O. E. Frivold and O. Hassel, Physik. Z. 24, 82.

1924

10 K. T. Compton and C. T. Zahn, Phys. Rev. 23, 781.
11 C. T. Zahn, ibid. 24, 400.

1925

12 J. Errera, J. phys. radium (6) 6, 390.
13 L. Lange, Z. Physik 33, 169.
14 C. P. Smyth and C. T. Zahn, J. Am. Chem. Soc. 47, 2501.

1926

15 J. Errera, C. r. acad. sci. 182, 1623.
16 "   "   Physik. Z. 27, 764.
17 K. Højendahl, Nature 117, 892.
18 W. Kliefoth, Z. Physik 39, 402.
19 R. Sänger, Physik. Z. 27, 165.
20 "   "   ibid. 27, 556.
21 C. T. Zahn, Phys. Rev. 27, 455.

1927

22 H-J von Braunmühl, Physik. Z. 28, 141.
23 L. Ebert and H. v. Hartel, Naturwiss. 15, 669.
24 I. J. Krchma and J. W. Williams, J. Am. Chem. Soc. 49, 2408.
25 C. V. Raman and K. S. Krishnan, Phil. Mag. (7) 3, 713.
26 R. Sänger, Physik. Z. 28, 455.
27 C. P. Smyth and S. O. Morgan, J. Am. Chem. Soc. 49, 1030.
28 A. M. Taylor and E. K. Rideal, Proc. Roy. Soc. (London) A 115, 589.
29 H. E. Watson, ibid. 117, 43.
30 J. W. Williams and R. J. Allgeier, J. Am. Chem. Soc. 49, 2416.
31 " "   "   and I. J. Krchma, ibid. 49, 1676.

<u>1927</u> (Cont'd.)

32. E. Wrede, Z. Physik <u>44</u>, 261.

<u>1928</u>

33 A. I. Anderson, Proc. Phys. Soc. (London) <u>40</u>, 62.
34 E. Bretscher, Helv. Phys. Acta <u>1</u>, 355.
35 L. Ebert, R. Eisenschitz, and H. v. Hartel, Z. physik. Chem. B <u>1</u>, 94.
36 J. Errera, J. phys. radium (6) <u>9</u>, 307.
37 "   "   Physik. Z. <u>29</u>, 426.
38 "   "   ibid. 29, 689.
39 "   "   Polarisation diélectrique, Paris; Leipziger Vorträge 1929, p. 105.
40 I. Estermann, Z. physik. Chem. B <u>1</u>, 134.
41 "   "   ibid. B <u>1</u>, 161.
42 "   "   ibid. B <u>1</u>, 422.
43 M. Forró, Z. Physik 47, 430.
44 W. Lautsch, Z. physik. Chem. B <u>1</u>, 115.
45 P. C. Mahanti and D. N. Sen-Gupta, Indian J. Physics <u>3</u>, 181.
46 " "   "   " " "   "   J. Indian Chem. Soc. <u>5</u>, 673.
47 J. Rolinski, Physik. Z. <u>29</u>, 658.
48 S. Rosental, Bull. intern. acad. Polonaise, A, p. 377.
49 R. Sänger and O. Steiger, Helv. Phys. Acta <u>1</u>, 369.
50 S. C. Sircar, Indian J. Phys. <u>3</u>, 197.
51 C. P. Smyth and S. O. Morgan, J. Am. Chem. Soc. <u>50</u>, 1547.
52 " "   "   " " "   "   and J. C. Boyce, ibid. <u>50</u>, 1536.
53 " "   "   and W. N. Stoops, ibid. <u>50</u>, 1883.
54 J. D. Stranathan, Phys. Rev. <u>31</u>, 156.
55 " "   "   ibid. <u>31</u>, 653.
56 H. A. Stuart, Z. Physik <u>47</u>, 457.
57 " "   "   ibid. <u>51</u>, 490.
58 J. W. Williams, J. Am. Chem. Soc. <u>50</u>, 2350.
59 " "   "   Physik. Z. <u>29</u>, 204.
60 " "   "   ibid. <u>29</u>, 271.
61 " "   "   ibid. <u>29</u>, 683.
62 " "   "   Z. physik. Chem. A <u>138</u>, 75.
63 " "   "   and E. F. Ogg, J. Am. Chem. Soc. <u>50</u>, 94.
64 " "   "   and C. H. Schwingel, ibid. <u>50</u>, 362.
65 " "   "   and A. Weissberger, ibid. <u>50</u>, 2332.
66 K. L. Wolf and E. Lederle, Physik. Z. <u>29</u>, 948; K. L. Wolf, Z. physik. Chem. B <u>2</u>, 39 (1929).
67 C. T. Zahn and J. B. Miles, Phys. Rev. <u>32</u>, 497.

<u>1929</u>

68 E. Bretscher, Helv. Phys. Acta <u>2</u>, 257.
69 "   "   and T. Wagner-Jauregg, ibid. <u>2</u>, 522.
70 R. J. Clark, Proc. Roy Soc. (London) A <u>124</u>, 689.
71 C. R. Daily, Phys. Rev. <u>34</u>, 548.
72 J. Errera and M. L. Sherrill, Leipziger Vorträge 1929, p. 41.
73 I. Estermann, Z. physik. Chem. B <u>2</u>, 287.
74 P. N. Ghosh, Nature <u>123</u>, 413.
75 " "   "   P. C. Mahanti, and B. C. Mukhergee, Z. Physik <u>58</u>, 200.
76 " " "   " "   "   and D. N. Sen-Gupta, ibid. <u>54</u>, 711.
77 O. Hassel and E. Naeshagen, Z. physik. Chem. B <u>4</u>, 217.
78 "   "   " "   "   ibid. B <u>6</u>, 152.
79 K. Højendahl, Physik. Z. <u>30</u>, 391.

1929 (Cont'd.)

80 P. C. Mahanti, J. Indian Chem. Soc. 6, 743.
81 "   "   "   and R. N. Das-Gupta, ibid. 6, 411.
82 "   "   "   "   "   "   "   Indian J. Physics 3, 467.
83 J. B. Miles, Jr., Phys. Rev. 34, 964.
84 A. Parts, Z. physik. Chem. B 4, 227.
85 R. Sänger, Leipziger Vorträge 1929, p. 1.
86 "   "   and O. Steiger, Helv. Phys. Acta 2, 136.
87 "   "   "   "   "   ibid. 2, 411.
88 R. Sängewald and A. Weissberger, Physik. Z. 30, 268.
89 C. P. Smyth, J. Am. Chem. Soc. 51, 2380.
90 "   "   "   and W. N. Stoops, ibid. 51, 3312.
91 "   "   "   "   "   "   ibid. 51, 3330.
92 P. Walden and O. Werner, Z. physik. Chem. B 2, 10.
93 A. Weissberger and R. Sängewald, Z. physik. Chem. B 5, 237.
94 "   "   "   "   "   Physik. Z. 30, 792.
95 "   "   and J. W. Williams, Z. physik. Chem. 3, 367.
96 O. Werner, Z. anorg. allgem. Chem. 181, 154.
97 "   "   Z physik. Chem. B 4, 371.
98 "   "   ibid. B 4, 393.
99 K. L. Wolf, Z. physik. Chem. B 3, 128.
100 "   "   "   and H. Volkmann, ibid. B 3, 139.

1930

101 E. Bergmann, L. Engel and S. Sándor, Ber. 63 B, 2572.
102 "   "   "   "   "   "   "   Z. physik. Chem. B 10, 106.
103 "   "   "   "   "   "   "   ibid. B 10, 397.
104 "   "   M. Magat and D. Wagenberg, Ber. 63 B, 2576.
105 G. Briegleb, Z. physik. Chem. B 10, 205.
106 L. M. Das and S. C. Roy, Indian J. Physics 5, 441.
107 D. Doborzynski, Z. Physik 66, 657; Bull. intern. acad. Polonaise A, p. 97.
108 H. L. Donle and G. Volkert, Z. physik. Chem. B 8, 60.
109 "   "   "   and K. L. Wolf, ibid. B 8, 55.
110 R. W. Dornte and C. P. Smyth, J. Am. Chem. Soc. 52, 3546.
111 A. E. Eide and O. Hassel, Tids. Kjemi Bergwesen 10, 93.
112 J. Errera and M. L. Sherrill, J. Am. Chem. Soc. 52, 1993.
113 K. Fredenhagen and F. Maske, Z. physik. Chem. B 10, 142.
114 O. Fuchs, Z. Physik 63, 824.
115 D. L. Hammick, R. G. A. New, N. V. Sidgwick and L. E. Sutton, J. Chem.
        Soc. 137, 1876.
116 O. Hassel, Z. Elektrochem. 36, 735.
117 "   "   and E. Naeshagen, Tids. Kjemi Bergwesen 10, 81.
118 "   "   "   "   "   ibid. 10, 128; Zentr. 1931, I p. 893; C. A.
        p. 1493.
119 O. Hassel and E. Naeshagen, Z. physik. Chem. B 6, 441.
120 "   "   "   "   "   ibid. B 8, 357.
121 "   "   and A. H. Uhl, ibid. B 8, 187.
122 "   "   "   "   "   Naturwiss. 18, 247.
123 F. G. Keyes and J. G. Kirkwood, Phys. Rev. 36, 1570.
124 P. C. Mahanti, Physik. Z. 31, 546.
125 L. Meyer, Z. physik. Chem. B 8, 27.
126 S. O. Morgan and H. H. Lowry, J. Phys. Chem. 34, 2385.
127 H. Müller and H. Sack, Physik. Z. 31, 815.
128 L. Orthner and G. Freyss, Ann. 484, 131.
129 N. N. Pal, Phil. Mag. (7) 10, 265.

1930 (Cont'd.)

130 A. Parts, Z. physik. Chem. B 7, 327.
131 "   "   ibid. B 10, 264.
132 S. Rosental, Z. Physik 66, 652.
133 R. Sänger, Helv. Phys. Acta 3, 161.
134 "   "   Physik. Z. 31, 306.
135 C. H. Schwingel and J. W. Williams, Phys. Rev. 35, 855.
136 D. N. Sen-Gupta, Nature 125, 600.
137 C. P. Smyth and H. E. Rogers, J. Am. Chem. Soc. 52, 1824.
138 " "   "   " " "   " ibid. 52, 2227.
139 O. Steiger, Helv. Phys. Acta 3, 161.
140 M. Velasco, Anales soc. españ. fís. quím. 28, 1228.
141 A. Weissberger in O. Hassel, Z. Elektrochem. 36, 735.
142 "   "   and R. Sängewald, Z. physik. Chem. B 9, 133.
143 J. W. Williams, J. Am. Chem. Soc. 52, 1831.
144 " "   "   ibid. 52, 1838.
145 " "   "   and J. M. Fogelberg, ibid. 52, 1356.
146 " "   "   " " "   Physik. Z. 31, 363.
147 K. L. Wolf, ibid. 31, 227.
148 " "   " Trans. Far. Soc. 26, 315.
149 C. T. Zahn, Phys. Rev. 35, 848.
150 " "   " ibid. 35, 1047.
151 C. Zakrzewski and D. Doborzyński, Bull. intern. acad. Polonaise A, p. 300.

1931

152 E. Bergmann and L. Engel, Physik. Z. 32, 507.
153 "   "   " "   " Z. Elektrochem. 37, 563.
154 "   "   " "   " Z. physik. Chem. B 13, 232.
155 "   "   " "   " ibid. B 15, 85.
156 "   "   and W. Schütz, Nature 128, 1077.
157 A. A. Bless, Phys. Rev. 37, 1149.
158 H. Braune and T. Asche, Z. physik. Chem. B 14, 18.
159 E. Bretscher, Physik. Z. 32, 765.
160 J. De Vries and W. H. Rodebush, J. Am. Chem. Soc. 53, 2888.
161 H. L. Donle, Z. physik. Chem. B 14, 326.
162 L. Ebert and K. Højendahl, ibid. B 15, 74.
163 J. M. Fogelberg and J. W. Williams, Physik. Z. 32, 27.
164 W. Gallay, Kolloid-Z. 57, 1.
165 P. N. Ghosh and T. P. Chatterjee, Phys. Rev. 37, 427.
166 P. Gross, Physik. Z. 32, 587.
167 O. Hassel, Z. Elektrochem. 37, 540.
168 "   "   and E. Naeshagen, Z. physik. Chem. B 12, 79.
169 "   "   "   "   "   ibid. B 14, 232.
170 E. C. E. Hunter and J. R. Partington, J. Chem. Soc. p. 2062.
171 H. Lütgert, Z. physik. Chem. B 14, 27.
172 "   "   ibid. B 14, 31.
173 "   "   ibid. B 14, 350.
174 N. Nakata, Ber. 64 B, 2059.
175 A. Parts, Z. physik. Chem. B 12, 312.
176 "   "   ibid. B 12, 323.
177 W. R. Pyle, Phys. Rev. 38, 1057.
178 R. Sänger, Physik. Z. 32, 21, 414.
179 J. H. Simons and G. Jessop, J. Am. Chem. Soc. 53, 1263.
180 C. P. Smyth and R. W. Dornte, ibid. 53, 545.
181 " "   " "   " "   " ibid. 53, 1296.

<u>1931</u> (Cont'd.)

182 C. P. Smyth and R. W. Dornte, ibid. <u>53</u>, 2005.
183 " " " " " " and E. B. Wilson, ibid. <u>53</u>, 4242.
184 " " " and S. E. Kamerling, ibid. <u>53</u>, 2988.
185 " " " and W. S. Walls, ibid. <u>53</u>, 527.
186 " " " " " " " ibid. <u>53</u>, 2115.
187 O. Steiger, Physik. Z. <u>32</u>, 425.
188 W. N. Stoops, J. Phys. Chem. <u>35</u>, 1704.
189 L. E. Sutton, Nature <u>128</u>, 639.
190 " " " Proc. Roy. Soc. (London) <u>133</u> A, 668.
191 " " " and T. W. J. Taylor, J. Chem. Soc. p. 2190.
192 L. Tiganik, Z. physik. Chem. B <u>13</u>, 425.
193 " " ibid. B <u>14</u>, 135.
194 H. Ulich and N. Nespital, Z. angew. Chem. <u>44</u>, 750. .
195 " " " " " Z. Elektrochem. <u>37</u>, 559.
196 E. Waldschmitt, Diss. Würzburg, 1930, in L. Ebert and K. Højendahl, Z. physik. Chem. B <u>15</u>, 74.
197 H. E. Watson, G. G. Rao and K. L. Ramaswamy, Proc. Roy. Soc. (London) A <u>132</u>, 569.
198 A. Weissberger, Z. physik. Chem. B <u>15</u>, 97.
199 " " and R. Sängewald, ibid. B <u>12</u>, 399.
200 " " " " ibid. B <u>13</u>, 383.
201 J. W. Williams and J. M. Fogelberg, J. Am. Chem. Soc. <u>53</u>, 2096.
202 K. L. Wolf and W. Bodenheimer, Z. physik. Chem., Bodenstein Festschr., p. 620.
203 K. L. Wolf and W. J. Gross, ibid. B <u>14</u>, 305.
204 " " " and H. G. Trieschmann, ibid. B <u>14</u>, 346.
205 J. Wyman, Jr., J. Biol. Chem. <u>90</u>, 443.
206 C. T. Zahn, Phys. Rev. <u>37</u>, 1516.
207 " " " ibid. <u>38</u>, 521.

<u>1932</u>

208 A. E. van Arkel and J. L. Snoek, Z. physik. Chem. B <u>18</u>, 159.
209 E. Bergmann, L. Engel and H. Hoffman, ibid. B <u>17</u>, 92.
210 " " " " " H. A. Wolff, ibid. B <u>17</u>, 81.
211 " " " " " H. Meyer, Ber. <u>65</u> B, 446.
212 " " and W. Schütz, Z. physik. Chem. B <u>19</u>, 389.
213 " " " " " ibid. B <u>19</u>, 395.
214 " " " " " ibid. B <u>19</u>, 401.
215 " " and M. Tschudnovsky, ibid. B <u>17</u>, 100.
216 " " " " " ibid. B <u>17</u>, 107.
217 " " " " " ibid. B <u>17</u>, 116.
218 " " " " " Ber. <u>65</u> B, 457.
219 W. Bodenheimer and K. Wehage, Z. physik. Chem. B <u>18</u>, 343.
220 G. Briegleb, ibid. B <u>16</u>, 276.
221 J. M. A. de Bruyne, R. M. Davis, and P. M. Gross, Physik. Z. <u>33</u>, 719.
222 S. Dobiński, Bull. intern. acad. Polonaise A p. 329.
223 H. L. Donle, Z. physik. Chem. B <u>18</u>, 146.
224 " " " and K. A. Gehrckens, ibid. B <u>18</u>, 316.
225 " " " " " " " in E. Müller, Ann. <u>495</u>, 132.
226 J. Errera, J. chim. phys. <u>29</u>, 577.
227 F. Fairbrother, J. Chem. Soc. p. 43.
228 W. Graffunder and E. Heymann, Z. physik. Chem. B <u>15</u>, 377.
229 E. W. Greene and J. W. Williams, Phys. Rev. <u>42</u>, 119.
230 E. Halmöy and O. Hassel, Z. physik. Chem. B <u>15</u>, 472.

1932 (Cont'd.)

231 D. L. Hammick, R. G. A. New, and L. E. Sutton, J. Chem. Soc. p. 742
232 O. Hassel and E. Naeshagen, Z. physik. Chem. B 15, 373.
233 "   "   "   "   "   ibid. B 15, 417.
234 "   "   "   "   "   ibid. B 19, 434.
235 L. Heil, Phys. Rev. 39, 666.
236 H. Hibbert and J. S. Allen, J. Am. Chem. Soc. 54, 4115.
237 K. Higashi, Bull. Inst. Phys.-Chem. Res. (Tokyo) 11, 729.
238 E. C. E. Hunter and J. R. Partington, J. Chem. Soc. p. 2812.
239 H. L. Knowles, J. Phys. Chem. 36, 2554.
240 R. J. W. Le Fèvre and J. W. Smith, J. Chem. Soc. p. 2239.
241 "   "   "   "   "   "   "   ibid. p. 2810.
242 E. P. Linton and O. Maass, Can. J. Res. 7, 81.
243 T. M. Lowry and J. Hofton, J. Chem. Soc. p. 207.
244 H. Lütgert, Z. physik. Chem. B 17, 460.
245 L. Meyer and A. Büchner, Physik. Z. 33, 390.
246 S. Mizushima and K. Higashi, Proc. Imp. Acad. (Tokyo) 8, 482.
247 W. Nespital, Z. physik. Chem. B 16, 153.
248 R. G. A. New and L. E. Sutton, J. Chem. Soc. p. 1415.
249 N. Nukada, Nia Kemio 5, 41.
250 W. Ostwald and R. Riedel, Koll.-Z. 59, 150.
251 A. Piekara, Phys. Rev. 42, 449; Acta phys. polon. 1, 393.
252 H. Poltz, O. Steil and O. Strasser, Z. physik. Chem. B 17, 155.
253 M. Puchalik, Physik. Z. 33, 341.
254 R. Sänger, O. Steiger, and K. Gächter, Helv. Phys. Acta 5, 200.
255 J. W. Smith, Proc. Roy. Soc. (London) A 136, 256.
256 "   "   "   ibid. A 138, 154.
257 "   "   "   and W. R. Angus, ibid. A 137, 372.
258 C. P. Smyth and W. S. Walls, J. Am. Chem. Soc. 54, 1854.
259 "   "   "   "   "   "   "   ibid. 54, 2261.
260 "   "   "   "   "   "   "   ibid. 54, 3230.
261 L. E. Sutton and J. B. Bentley, Nature 130, 314.
262 H. Ulich, E. Hertel and W. Nespital, Z. physik. Chem. B 17, 21.
263 "   "   "   "   "   "   "   ibid. B 17, 369.
264 A. Weissberger and H. Bach, Ber. 65 B, 24.
265 "   "   and R. Sängewald, ibid. 65 B, 701.
266 A. H. White and S. O. Morgan, Physics 2, 313.
267 A. J. Wildschut, Physica 12, 194.
268 E. Wolf, Z. physik. Chem. B 17, 46.
269 C. T. Zahn, Phys. Rev. 40, 291.
270 "   "   "   Physik. Z. 33, 525.
271 "   "   "   ibid. 33, 686.
272 "   "   "   ibid. 33, 730.

1933

273 A. E. van Arkel, Rec. trav. chim. 52, 733.
274 "   "   "   "   and J. L. Snoek, ibid. 52, 719.
275 L. F. Audrieth, W. Nespital and H. Ulich, J. Am. Chem. Soc. 55, 673.
276 J. M. A. de Bruyne, R. M. Davis, and P. M. Gross, ibid. 55, 3936.
277 L. Cambi, Gazz. chim. ital. 63, 579.
278 K. S. Chang and Y-T. Cha, J. Chinese Chem. Soc. 1, 107.
279 E. G. Cowley and J. R. Partington, J. Chem. Soc. p. 1252.
280 "   "   "   "   "   "   "   ibid. p. 1255.
281 "   "   "   "   "   "   "   ibid. p. 1257.
282 "   "   "   "   "   "   "   ibid. p. 1259.

1933 (Cont'd.)

283 H. J. Curtis, J. Chem. Physics 1, 160.
284 G. Devoto. Gazz. chim. ital. 63, 495.
285 S. Dobiński, Z. Physik 83, 129.
286 I. Estermann and R. G. J. Fraser, J. Chem. Physics 1, 390.
287 "   "   and M. Wohlwill, Z. physik. Chem. B 20, 195.
288 F. Fairbrother, J. Chem. Soc. p. 1541.
289 "   "   Proc. Roy. Soc. (London) A 142, 173.
290 E. H. Farmer and N. J. H. Wallis, J. Chem. Soc. p. 1304.
291 "   "   and F. L. Warren, ibid. p. 1297.
292 "   "   "   " "   " ibid. p. 1302.
293 O. Fuchs and H. L. Donle, Z. physik. Chem. B 22, 1.
294 K. A. Gehrckens and E. Müller, Ann. 500, 296.
295 F. R. Goss, J. Chem. Soc. p. 1341.
296 G. C. Hampson, R. H. Farmer, and L. E. Sutton, Proc. Roy. Soc. (London)
        A 143, 147.
297 P. C. Henriquez, Physica 1, 41.
298 K. Higashi, Bull. Inst. Phys.-Chem. Res. (Tokyo) 12, 22.
299 "   "   ibid. 12, 771.
300 "   "   ibid. 12, 780.
301 C. Hrynakowski and C. Kalinowski, C. r. acad. sci. 197, 483.
302 G. C. E. Hunter and J. R. Partington, J. Chem. Soc. p. 87.
303 "   "   "   "   "   "   " ibid. p. 309.
304 K. Kalinowski, Roczniki Chem. 13, 384.
305 S. E. Kamerling and C. P. Smyth, J. Am. Chem. Soc. 55, 462.
306 C. A. Kraus and G. S. Hooper, Proc. Nat. Acad. Sci. 19, 939.
307 K. F. Luft, Z. Physik 84, 767.
308 K. B. McAlpine and C. P. Smyth, J. Am. Chem. Soc. 55, 453.
309 J. G. Malone, J. Chem. Phys. 1, 197.
310 G. S. Parsons and C. W. Porter, J. Am. Chem. Soc. 55, 4745.
311 J. N. Pearce and L. Berhenke, Proc. Iowa Acad. Sci. 40, 93.
312 A. Piekara, Bull. intern. acad. polonaise A p. 333.
313 H. Poltz, Z. physik. Chem. B 20, 351.
314 M. Puchalik, Acta Phys. Polonica 2, 305.
315 I. Sakurada and M. Taniguchi, Bull. Inst. Phys.-Chem. Res. (Tokyo) 12, 224.
316 N. V. Sidgwick, L. E. Sutton, and W. Thomas, J. Chem. Soc. p. 406.
317 J. W. Smith, ibid. p. 1567.
318 C. P. Smyth and K. B. McAlpine, J. Chem. Physics 1, 60.
319 "   "   "   "   "   " ibid. 1, 190.
320 "   "   1, 200.
321 M. E. Spaght, F. Hein and H. Pauling, Physik. Z. 34, 212.
322 L. E. Sutton, R. G. A. New and J. B. Bentley, J. Chem. Soc. p. 652.
323 T. W. J. Taylor and L. E. Sutton, ibid. p. 63.
324 W. Theilacker, Z. physik. Chem. B 20, 142.
325 H. H. Uhlig, J. G. Kirkwood and F. G. Keyes, J. Chem. Physics 1, 155.
326 W. S. Walls and C. P. Smyth, J. Chem Physics 1, 337.
327 A. Weissberger and R. Sängewald, Z. physik. Chem. B 20, 145.
328 M. Wohlwill, Z. Physik 80, 67.
329 K. L. Wolf and O. Strasser, Z. physik. Chem. B 21, 389.
330 J. Wyman, Jr. and T. L. McMeekin, J. Am. Chem. Soc. 55, 915.
331 C. T. Zahn, Physik. Z. 34, 461.
332 "   "   " ibid. 34, 570.

1934

333 J. S. Allen and H. Hibbert, J. Am. Chem. Soc. 56, 1398.

1934 (Cont'd.)

334 A. E. van Arkel and J. L. Snoek, Physik. Z. 35, 187.
335 "   "   "   "   " "   "   "   Rec. trav. chim. 53, 91.
336 "   "   "   "   " "   "   Trans. Far. Soc. 30, 707.
337 F. Barrow and F. J. Thorneycroft, J. Chem. Soc. p. 722.
338 G. M. Bennett, D. P. Earp and S. Glasstone, ibid. p. 1179.
339 "   "   "   and S. Glasstone, ibid. p. 128.
340 G. Briegleb and J. Kambeitz, Naturwiss. 22, 105.
341 "   "   "   " "   "   Z. physik. Chem. B 25, 251.
342 "   "   "   "   "   ibid. B 27, 11.
343 F. Brown, J. M. A. de Bruyne and P. M. Gross, J. Am. Chem. Soc. 56, 1291.
344 E. Czerlinsky, Z. Physik 88, 515.
345 F. Fairbrother, J. Chem. Soc. p. 1846.
346 "   "   Nature 134, 458.
347 "   "   Trans. Far. Soc. 30, 862.
348 F. R. Goss, J. Chem. Soc. p. 696.
349 " "   " ibid. p. 1467.
350 L. G. Groves and S. Sugden, ibid. p. 1094.
351 D. L. Hammick, R. G. A. New, and R. B. Williams, ibid. p. 29.
352 G. C. Hampson, Trans. Far. Soc. 30, 877.
353 H. Harms, Z. physik. Chem. B 30, 440.
354 P. C. Henriquez, Rec. trav. chim. 53, 1139.
355 K. Higashi, Bull. Inst. Phys.-Chem. Res. (Tokyo) 13, 186.
356 "   "   ibid. 13, 703.
357 "   "   ibid. 13, 1167.
358 "   "   Sci. Papers Inst. Phys.-Chem. Res. (Tokyo) 24, 57.
359 G. S. Hooper and C. A. Kraus, J. Am. Chem. Soc. 56, 2265.
360 H. O. Jenkins, J. Chem. Soc. p. 480.
361 " "   "   Nature 133, 106.
362 " "   "   Trans. Far. Soc. 30, 739.
363 W. D. Kumler and C. W. Porter, J. Am. Chem. Soc. 56, 2549.
364 E. Landt, Naturwiss. 22, 809.
365 M. G. Malone and A. L. Ferguson, J. Chem. Physics 2, 99.
366 M. Milone, IX Congr. intern. quim. pura applicada 2, 191; Gazz. chim. ital.
       65, 94 (1935).
367 F. H. Müller, Physik. Z. 35, 1009.
368 H. Müller, ibid. 35, 346.
369 E. Naeshagen, Z. physik. Chem. B 25, 157.
370 "   "   in A. Langseth and B. Qviller, ibid. B 27, 79.
371 W. Ostwald and R. Riedel, Koll.-Z. 69, 185.
372 M. M. Otto and H. H. Wenzke, Ind. Eng. Chem. Anal. Ed. 6, 187.
373 " "   " "   "   "   J. Am. Chem. Soc. 56, 1314.
374 J. R. Partington, Trans. Far. Soc. 30, 822.
375 J. N. Pearce and J. Berhenke, Proc. Iowa Acad. Sci. 41, 141.
376 D. J. Pflaum and H. H. Wenzke, J. Am. Chem. Soc. 56, 1106.
377 A. Piekara and B. Piekara, C. r. acad. sci. 198, 1018.
378 M. Puchalik, Acta Phys. Polonica 3, 179.
379 M. A. G. Rau and B. N. Narayanaswamy, Proc. Indian Acad. Soc. 1 A, 14.
380 "   " "   "   "   "   "   "   ibid. 1 A, 217.
381 "   " "   "   "   " "   "   ibid. 1 A, 489.
382 "   " "   "   "   " "   "   Z. physik. Chem. B 26, 23.
383 I. Sakurada and S. Lee, J. Soc. Chem. Ind. Japan 37 Suppl., 331.
384 H. Scheffers, Physik. Z. 35, 425.
385 C. H. Schwingel and E. W. Greene, J. Am. Chem. Soc. 56, 653.
386 M. L. Sherrill, M. E. Smith and D. D. Thompson, ibid. 56, 611.
387 C. P. Smyth and K. B. McAlpine, ibid. 56, 1697.

BIBLIOGRAPHY

<u>1934</u> (Cont'd.)

388 C. P. Smyth and K. B. McAlpine, J. Chem. Physics <u>2</u>, 499.
389 " " " " " " " ibid. <u>2</u>, 571.
390 J. L. Snoek, Physik. Z. <u>35</u>, 196.
391 A. E. Stearn and C. P. Smyth, J. Am. Chem. Soc. <u>56</u>, 1667.
392 S. Sugden, Nature <u>133</u>, 415.
393 L. E. Sutton, Trans. Far. Soc. <u>30</u>, 789.
394 H. E. Watson, G. G. Rao and K. L. Ramaswamy, Proc. Roy. Soc. (London)
        A <u>143</u>, 558.
395 A. Weissberger, R. Sängewald and G. C. Hampson, Trans. Far. Soc. <u>30</u>,
        884.
396 H. H. Wenzke and R. P. Allard, J. Am. Chem. Soc. <u>56</u>, 858.
397 J. W. Williams, C. H. Schwingel and C. H. Winning, ibid. <u>56</u>, 1427.
398 C. J. Wilson and H. H. Wenzke, ibid. <u>56</u>, 2025.
399 " " " " " " " " J. Chem. Physics <u>2</u>, 546.

<u>1935</u>

400 P. Abadie and G. Champetier, C. r. acad. sci. <u>200</u>, 1590.
401 W. R. Angus, A. H. Leckie, C. G. Le Fèvre, R. J. W. Le Fèvre and
        A. Wassermann, J. Chem. Soc. p. 1751.
402 F. Arndt, G. T. O. Martin and J. R. Partington, ibid. p. 602.
403 E. Bergmann, ibid. p. 987.
404 " " and G. C. Hampson, ibid. p. 989.
405 " " " A. Weizmann, J. Am. Chem. Soc. <u>57</u>, 1755.
406 J. Böeseken, F. Tellegen and P. C. Henriquez, Rec. trav. chim. <u>54</u>, 733.
407 J. M. A. de Bruyne and C. P. Smyth, J. Am. Chem. Soc, <u>57</u>, 1203.
408 A. Burawoy and C. S. Gibson, J. Chem. Soc. p. 219.
409 H. J. Cavell and S. Sugden, ibid. p. 621.
410 F. Challenger and J. B. Harrison, J. Inst. Petroleum Tech. <u>21</u>, 135.
411 E. G. Cowley and J. R. Partington, J. Chem. Soc. p. 604.
412 " " " " " " " Nature <u>136</u>, 643.
413 W. J. Curran and H. H. Wenzke, J. Am. Chem. Soc. <u>57</u>, 2162.
414 J. T. Edsall and J. Wyman, Jr., ibid. <u>57</u>, 1964.
415 C. S. Gibson, Sci. J. Roy. Coll. Sci. <u>5</u>, 54.
416 C. A. Goethals, Rec. trav. chim. <u>54</u>, 299.
417 L. G. Groves and S. Sugden, J. Chem. Soc. p. 971.
418 D. L. Hammick, G. C. Hampson and G. I. Jenkins, Nature <u>136</u>, 990; J. Chem.
        Soc. 1938 p. 1263.
419 C. Hennings, Z. physik. Chem. B <u>28</u>, 267.
420 E. Hertel and E. Dumont, ibid. B <u>30</u>, 139.
421 A. Jagielski and J. Weslowski, Bull. intern. acad. Polonaise A p. 260.
422 K. A. Jensen, Z. anorgan. allgem. Chem. <u>225</u>, 97.
423 M. Kubo, Sci. Papers Inst. Phys.-Chem. Res. (Tokyo) <u>26</u>, 242.
424 " " ibid. <u>27</u>, 65.
425 S. Kambara, J. Soc. Chem. Ind. Japan <u>38</u> Suppl. 10, 506; Rubber Chem.
        Tech. <u>9</u>, 296 (1936); India-Rubber J. <u>92</u>, 24 (1936).
426 C. G. Le Fèvre and R. J. W. Le Fèvre, J. Chem. Soc. p. 957.
427 " " " " " " " " " " ibid. p. 1696.
428 " " " " " " " " " " and K. W. Robertson, ibid. p. 480.
429 P. C. Mahanti, Phil. Mag. (7) <u>20</u>, 274.
430 " " " Z. Physik <u>94</u>, 220.
431 A. R. Martin, Nature <u>135</u>, 909.
432 K. B. McAlpine and C. P. Smyth, J. Chem. Physics <u>3</u>, 55.
433 M. Milone, Gazz. chim. ital. <u>65</u>, 152.
434 S. Mizushima, K. Suenaga and K. Kojima, Bull. Chem. Soc. Japan <u>10</u>, 167.

1935 (Cont'd.)

435 N. Nakata, Bull. Chem. Soc. Japan, 10, 31'8.
436 W. Ostwald and R. Riedel, Koll.-Z. 70, 75.
437 M. M. Otto, J. Am. Chem. Soc. 57, 693.
438 " " " ibid. 57, 1476.
439 " " " and H. H. Wenzke, ibid. 57, 294.
440 G. R. Paranjpe and P. Y. Deshpande, Proc. Indian Acad. Sci. 1 A, 880.
441 J. R. Partington and E. G. Cowley, Nature 135, 474.
442 " " " " " " ibid. 135, 1038.
443 J. N. Pearce and L. F. Berhenke, J. Phys. Chem. 39, 1005.
444 M. Puchalik, Acta Phys. Polon. 4, 145.
445 K. L. Ramaswamy, Proc. Indian Acad. Sci. 2 A, 364.
446 M. A. G. Rau and S. S. Rao, ibid. 2 A, 232.
447 R. Rollefson and A. H. Rollefson, Phys. Rev. 48, 779.
448 I. Sakurada and S. Lee, Koll.-Z. 72, 320.
449 C. P. Smyth and K. B. McAlpine, J. Am. Chem. Soc. 57, 979.
450 " " " " " J. Chem. Physics 3, 347.
451 " " " and W. S. Walls, ibid. 3, 557.
452 D. Sundhoff and H-J. Schumacher, Z. physik. Chem. B 28, 17.
453 L. E. Sutton and G. C. Hampson, Trans. Far. Soc. 31, 945.
454 W. J. Svirbely, J. E. Ablard and J. C. Warner, J. Am. Chem. Soc. 57, 652.
455 P. Trunel, C. r. acad. sci. 200, 557,
456 " " ibid. 200, 2186.
457 H. Ulich, H. Peisker and L. F. Audrieth, Ber. 68 B, 1677.
458 W. Wassiliew, J. Syrkin and I. Kenez, Nature 135, 71.
459 A. Weissberger and R. Sängewald, J. Chem. Soc. p, 855.
460 C. J. Wilson and H. H. Wenzke, J. Am. Chem. Soc. 57, 1265.

1936

461 E. Bergmann, J. Chem. Soc. p. 402.
462 " " and J. Hirshberg, ibid. p. 331.
463 " " " A. Weizmann, Trans. Far. Soc. 32, 1318.
464 " " " " ibid. 32, 1327.
465 G. N. Bhattacharyya, Indian J. Physics 10, 281.
466 J. Böeseken, P. C. Henriquez and J. J. van der Spek, Rec. trav. chim. 55, 145.
467 E. G. Cowley and J. R. Partington, J. Chem. Soc. p. 45.
468 " " " " " " " ibid. p. 47.
469 " " " " " " ibid. p. 1184.
470 F. Fairbrother, ibid. p. 847.
471 F. C. Frank, ibid. p. 1324.
472 J. A. Geddes and C. A. Kraus, Trans. Far. Soc. 32, 585.
473 G. C. Hampson and A. Weissberger, J. Am. Chem. Soc. 58, 2111.
474 " " " " " J. Chem. Soc. p. 393.
475 K. Higashi, Bull. Inst. Phys.-Chem. Res. (Tokyo) 15, 766.
476 " " Sci. Papers Inst. Phys.-Chem. Res. (Tokyo) 28, 284.
477 F. E. Hoecker, J. Chem. Physics 4, 431.
478 T-Y. Hsü and C-T. Kwei, J. Chinese Chem. Soc. 4, 105.
479 H. O. Jenkins, J. Chem. Soc. p. 862.
480 " " " ibid. p. 1049.
481 K. A. Jensen, Z. anorg. allgem. Chem. 229, 225.
482 " " " ibid. 229, 265.
483 " " " ibid. 229, 282.
484 S. Kambara, J. Soc. Chem. Ind. Japan 39, Suppl. 4, 138; India Rubber J. 93, 247 (1937).

<u>1936</u> (Cont'd.)

485 M. Kubo, Sci. Papers Inst. Phys.-Chem. Res. (Tokyo) <u>29</u>, 122.
486 "   "   ibid. <u>29</u>, 179.
487 "   "   ibid. <u>30</u>, 169.
488 "   "   ibid. <u>30</u>, 238.
489 W. D. Kumler, J. Am. Chem. Soc. <u>58</u>, 1049.
490 C. G. Le Fèvre and R. J. W. Le Fèvre, J. Chem. Soc. p. 398.
491 " " " "   " " " "   "   "   ibid. p. 487.
492 " " " "   " " " "   "   "   ibid. p. 1130.
493 R. J. W. Le Fèvre and P. Russell, ibid. p. 491.
494 " " " "   "   " "   "   ibid. p. 496.
495 R. J. B. Marsden and L. E. Sutton, ibid. p. 599.
496 " " "   "   " " "   "   ibid. p. 1383.
497 G. Martin, Physik. Z. 37, 665.
498 G. T. O. Martin and J. R. Partington, J. Chem. Soc. p. 158.
499 " " "   "   " " "   "   ibid. p. 1175.
500 " " "   "   " " "   "   ibid. p. 1178.
501 " " "   "   " " "   "   ibid. p. 1182.
502 S. Mizushima, Y. Morino and K. Kojima, Sci. Papers Inst. Phys.-Chem.
        Res. (Tokyo) <u>29</u>, 111.
503 H. Moureu, C. r. acad. sci. <u>202</u>, 314.
504 A. Niini, Ann. Acad. Sci. Fennicae A <u>46</u>, No. 1.
505 M. A. G. Rau, Current Sci. <u>5</u>, 132.
506 " " "   "   Proc. Indian Acad. Sci. <u>4</u> A, 687.
507 N. V. Sidgwick and H. D. Springall, J. Chem. Soc. p. 1532.
508 C. P. Smyth and S. A. McNeight, J. Am. Chem. Soc. <u>58</u>, 1723.
509 H. G. Trieschmann, Z. physik. Chem. B <u>33</u>, 283.
510 P. Trunel, C. r. acad. sci. 202, 37.
511 "   "   ibid. <u>203</u>, 563.
512 C-L. Tseng, C. Liu and C. E. Sun, J. Chinese Chem. Soc. <u>4</u>, 473.
513 H. E. Watson and K. L. Ramaswamy, Proc. Roy. Soc. (London) A <u>156</u>, 130.
514 " "   " , G. P. Kane and K. L. Ramaswamy, ibid. A <u>156</u>, 137.
515 J. W. Williams, C. H. Schwingel and C. H. Winning, J. Am. Chem. Soc. <u>58</u>,
        197.
516 J. Wyman, Jr., Chem. Rev. <u>19</u>, 213.

<u>1937</u>

517 S. Arrhenius, J. Chem. Physics <u>5</u>, 63.
518 M. Beyaert, Natuurw. Tijdschr. <u>19</u>, 197.
519 R. H. Birtles and G. C. Hampson, J. Chem. Soc. p. 10.
520 J. Böeseken and H. V. Takes, Rec. trav. chim. <u>56</u>, 858.
521 R. C. L. Bosworth, Proc. Cambridge Phil. Soc. <u>33</u>, 394.
522 " " "   "   and E. K. Rideal, Proc. Roy. Soc. (London) A <u>162</u>, 1.
523 W. B. Bridgman and J. W. Williams, J. Am. Chem. Soc. <u>59</u>, 1579; W. B.
        Bridgman, ibid. <u>60</u>, 530 (1938).
524 E. Briner, E. Perrottet, H. Paillard and B. Susz, Helv. Chim. Acta <u>20</u>, 762.
525 A. Burawoy, C. S. Gibson, G. C. Hampson and H. M. Powell, J. Chem. Soc.
        p. 1690.
526 S. L. Chien and T. C. Lay, J. Chinese Chem. Soc. <u>5</u>, 204.
527 E. G. Cowley and J. R. Partington, J. Chem. Soc. p. 130.
528 " " "   " " "   "   Nature <u>140</u>, 1100; J. Chem. Soc. 1938,
        p. 977.
529 B. C. Curran and H. H. Wenzke, J. Am. Chem. Soc. <u>59</u>, 943.
530 F. Eisenlohr and L. Hill, Z. physik. Chem. B <u>36</u>, 30.
531 "   "   " A. Metzner, ibid. A <u>178</u>, 350.

1937 (Cont'd.)

532 P. R. Frey and E. C. Gilbert, J. Am. Chem. Soc. 59, 1344.
533 H. L. Goebel and H. H. Wenzke, ibid. 59, 2301.
534 F. R. Goss, J. Chem. Soc. p. 1915.
535 U. Grassi and L. Puccianti, Nuovo cimento 14, 461.
536 L. G. Groves and S. Sugden, J. Chem. Soc. p. 158.
537 "   "   "   " "   "   ibid. p. 1779.
538 "   "   "   " "   "   ibid. p. 1782.
539 K. Higashi, Sci. Papers Inst. Phys.-Chem. Res. (Tokyo) 31, 311.
540 "   "   ibid. 31, 317.
541 F. E. Hoecker, J. Chem. Physics 5, 372.
542 E. D. Hughes, C. G. Le Fèvre and R. J. W. Le Fèvre, J. Chem. Soc. p. 202.
543 A. Jagielski, Bull. intern. acad. polon. sci., Classe sci. math. nat. A, 312.
544 K. A. Jensen, Z. anorg. allgem. Chem. 231, 365.
545 S. M. Koehl and H. H. Wenzke, J. Am. Chem. Soc. 59, 1418.
546 K. Kojima and S. Mizushima, Sci. Papers Inst. Phys.-Chem. Res. (Tokyo)
        31, No. 697, 296.
547 M. Kubo, ibid. 32, 26.
548 "   "   Y. Morino and S. Mizushima, ibid. 32, 129.
549 K. Lauer, Ber. 70 B, 1127.
550 C. E. Leberknight and J. A. Ord, Phys. Rev. 51, 430.
551 C. G. Le Fèvre and R. J. W. Le Fèvre, J. Chem. Soc. p. 196.
552 " " "   "   " " "   "   "   ibid. p. 1088.
553 R. J. W. Le Fèvre and H. Vine, ibid. p. 1805.
554 " " "   "   "   "   "   "   Chem. and Ind. p. 688.
555 R. Linke and W. Rohrmann, Z. physik. Chem. B 35, 256.
556 S. Mizushima, Y. Morino and M. Kubo, Physik. Z. 38, 459.
557 H. Mohler and J. Sorge, Helv. Chim. Acta 20, 1447.
558 A. Müller, Proc. Roy. Soc. (London) A 158, 403.
559 F. H. Müller, Physik. Z. 38, 283.
560 M. M. Otto, J. Am. Chem. Soc. 59, 1590.
561 J. R. Partington and D. I. Coomber, Nature 139, 510.
562 J. T. Randall, Proc. Roy. Soc. (London) A 159, 83.
563 M. A. G. Rau and N. Anantanarayanan, Proc. Indian Acad. Sci. 5 A, 185.
564 V. I. Romanov and I. A. Elstin, Physik. Z. Sowjetunion 11, 526.
565 A. Smits, N. F. Moerman and J. C. Pathius, Z. physik. Chem. B 35, 60.
566 J. D. Stranathan, J. Chem. Physics 5, 828.
567 C. E. Sun and C. Liu, J. Chinese Chem. Soc. 5, 39.
568 P. Trautteur, Nuovo cimento 14, 265.
569 P. Trunel, C. r. acad. sci. 205, 236.
570 C-L. Tseng, C. E. Sun and C. H. Yao, J. Chinese Chem. Soc. 5, 236.
571 P. Tuomikoski and A. Niini, Ann. Acad. Sci. Fennicae A 48, No. 11.
572 F. L. Warren, J. Chem. Soc. p. 1858.
573 W. G. Wassiliew and J. K. Syrkin, Acta Physicochim. U.R.S.S. 6, 639.
574 C-H. Yao and C. E. Sun, J. Chinese Chem. Soc. 5, 22.

1938

575 R. P. Bell and I. E. Coop, Trans. Far. Soc. 34, 1209.
576 E. Bergmann and A. Weizmann, Chem. and Ind. p. 364.
577 "   "   "   "   "   J. Am. Chem. Soc. 60, 1801.
578 "   "   "   "   "   Trans. Far. Soc. 34, 783.
579 M. Beyaert and F. Govaert, Natuurw. Tijdschr. 20, 119.
580 E. Briner, D. Frank and E. Perrottet, Helv. Chim. Acta 21, 1312.
581 L. O. Brockway and I. E. Coop, Trans. Far. Soc. 34, 1429.
582 I. G. M. Campbell, C. G. Le Fèvre, R. J. W. Le Fèvre and E. E. Turner,

76

BIBLIOGRAPHY

<u>1938</u> (Cont'd.)

J. Chem. Soc. p. 404.
583   D. I. Coomber and J. R. Partington, ibid. p. 1444.
584   I. E. Coop and L. E. Sutton, ibid. p. 1869.
585   E. G. Cowley and J. R. Partington, ibid. p. 1598.
586   J. D. Ferry and J. L. Oncley, J. Am. Chem. Soc. <u>60</u>, 1123.
587   R. Fonteyne, Natuurw. Tijdschr. <u>20</u>, 275.
588   V. de Gaouck and R. J. W. Le Fèvre, J. Chem. Soc. p. 741.
589   H. L. Goebel and H. H. Wenzke, J. Am. Chem. Soc. <u>60</u>, 697.
590   M. Gorman, R. M. Davis and P. M. Gross, Physik. Z. <u>39</u>, 181.
591   D. L. Hammick and R. B. Williams, J. Chem. Soc. p. 211.
592   G. S. Hartley, ibid. p. 633.
593   K. Higashi, Bull. Chem. Soc. Japan <u>13</u>, 158.
594   J. A. C. Hugill, I. E. Coop and L. E. Sutton, Trans. Far. Soc. <u>34</u>, 1518.
595   K. A. Jensen and B. Bak, J. prakt. Chem. <u>151</u>, 167.
596   F. J. Krieger and H. H. Wenzke, J. Am. Chem. Soc. <u>60</u>, 2115.
597   S. Lee and I. Sakurada, Koll.-Z. <u>82</u>, 72.
598   R. J. W. Le Fèvre, Dipole Moments, Methuen & Co. Ltd., London, pp. 73
        and 93.
599   R. J. W. Le Fèvre and G. J. Rayner, J. Chem. Soc. p. 1921.
600   "   "   "   "   "   "   H. Vine, ibid. p. 431.
601   "   "   "   "   "   "   "   "   ibid. p. 967.
602   "   "   "   "   "   "   "   "   ibid. p. 1790.
603   "   "   "   "   "   "   "   "   ibid. p. 1795.
604   "   "   "   "   "   "   "   "   ibid. p. 1878.
605   F. E. Lindquist and C. L. A. Schmidt, Compt. rend. trav. Lab. Carlsberg,
        Sér. chim. <u>22</u>, 307.
606   B. A. Middleton and J. R. Partington, Nature <u>141</u>, 516.
607   M. Milone and G. Tappi, Atti X cong. intern. chim. <u>2</u>, 352.
608   S. Mizushima, Y. Morino and H. Okazaki, Sci. Papers Inst. Phys.-Chem.
        Res. (Tokyo) <u>34</u>, 1147.
609   H. Mohler, Helv. Chim. Acta <u>21</u>, 67.
610   "   "   ibid. <u>21</u>, 784.
611   "   "   ibid. <u>21</u>, 787.
612   "   "   ibid. <u>21</u>, 789.
613   J. L. Oncley, J. Am. Chem. Soc. <u>60</u>, 1115.
614   G. R. Paranjpe and D. J. Davar, Indian J. Physics <u>12</u>, 283.
615   J. R. Partington and D. I. Coomber, Nature <u>141</u>, 918.
616   V. A. Plotnikov, I. A. Sheka, Z. A. Yankelevich, Mem. Inst. Chem., Acad.
        Sci. Ukrain S.S.R. <u>4</u>, 382.
617   A. Riedinger, Physik. Z. <u>39</u>, 380.
618   I. Sakurada and S. Lee, Koll.-Z. <u>82</u>, 67.
619   G. Scheibe and O. Stoll, Ber. <u>71</u> B, 1571.
620   E. A. Shott-L'vova and J. K. Syrkin, J. Phys. Chem. (U.S.S.R.) <u>12</u>, 479.
621   R. W. Schulz, Z. Physik <u>109</u>, 517.
622   T. Voitila, Ann. Acad. Sci. Fennicae A <u>49</u>, No. 1.
623   F. L. Warren, J. Chem. Soc. p. 1100.
624   W. G. Wassiliew and J. K. Syrkin, Acta Physicochim. U.R.S.S. <u>9</u>, 203.
625   "   "   "   " "   "   "   J. Phys. Chem. (U.S.S.R.) <u>12</u>, 153.
626   W. West and R. B. Killingsworth, J. Chem. Physics <u>6</u>, 1.

<u>1939</u>

627   J. W. Baker and L. G. Groves, J. Chem. Soc. p. 1144.
628   F. Barrow and F. J. Thorneycroft, ibid. p. 773.
629   J. Benoit and G. Ney, C. r. acad. sci. <u>208</u>, 1888.

1939 (Cont'd.)

630 E. Bergmann and A. Weizmann, J. Am. Chem. Soc. 61, 3583.
631 E. Briner, A. Gelbert and F. Perrottet, Helv. Chim. Acta 22, 1491.
632 "   "   K. Ryffel and E. Perrottet, ibid. 22, 927
633 W. H. Byers, J. Chem. Physics 7, 175.
634 C. C. Caldwell and R. J. W. Le Fèvre, J. Chem. Soc. p. 1614.
635 " "   "   " " " " " " Nature 143, 803.
636 I. E. Coop and L. E. Sutton, Trans. Far. Soc. 35, 505.
637 D. J. Davar, Current Sci. 8, 414.
638 I. Dostrovsky and R. J. W. Le Fèvre, J. Chem. Soc. p. 535.
639 M. A. Elliott and S. F. Acree, J. Res. U. S. Nat'l. Bur. Stand. 23, 675.
640 E. Fischer and F. Rogowski, Physik. Z. 40, 331.
641 V. de Gaouck and R. J. W. Le Fèvre, J. Chem. Soc. p. 1392.
642 " "   "   " " " " " ibid. p. 1457.
643 E. N. Gur'yanova and J. K. Syrkin, Acta Physicochim. U.R.S.S. 11, 657.
644 E. Halmöy and O. Hassel, J. Am. Chem. Soc. 61, 1601.
645 G. S. Hartley and R. J. W. Le Fèvre, J. Chem. Soc. p. 531.
646 J. Henrion, Bull. soc. roy. sci. Liége 8, 36.
647 K. Higashi and S. Uyeo, Bull. Chem. Soc. Japan 14, 87.
648 C. E. Ingham and G. C. Hampson, J. Chem. Soc. p. 981.
649 S. Kambara, J. Soc. Chem. Ind., Japan 42 Suppl., 314.
650 R. J. W. Le Fèvre and C. A. Parker, J. Chem. Soc. p. 677.
651 G. L. Lewis and C. P. Smyth, J. Am. Chem. Soc. 61, 3063.
652 " "   "   " " "   " ibid. 61, 3067.
653 " "   "   " " "   " J. Chem. Physics 7, 1085.
654 N. C. C. Li, ibid. 7, 1068.
655 B. K. Maïbaum, J. Exptl. Theoret. Phys. (U.R.S.S.) 9, 1383.
656 L. Malatesta, Gazz. chim. ital. 69, 629.
657 G. P. Mikhailov and D. V. Tishchenko, J. Gen. Chem. (U.S.S.R.) 9, 782.
658 N. L. Phalnikar and B. V. Bhide, Current Sci. 8, 473.
659 V. A. Plotnikov, I. A. Sheka and Z. A. Yankelevich, J. Gen. Chem.
        (U.S.S.R.) 9, 868.
660 H. de V. Robles, Rec. trav. chim. 58, 111.
661 E. A. Shott-L'vova and J. K. Syrkin, Acta Physicochim. U.R.S.S. 11, 659.
662 A. Skita and W. Faust, Ber. 72 B, 1127.
663 "   "   " R. Rössler, ibid. 72 B, 265.
664 B. Tamamushi, H. Akiyama and S. Umerzawa, Bull. Chem. Soc. Japan 14,
        310.
665 B. Tamamushi, H. Akiyama and S. Umerzawa, ibid. 14, 318.
666 G. Tappi and U. di Vajo, Gazz. chim. ital. 69, 615.
667 W. Theilacker and K. Fauser, Ann. 539, 103.
668 G. Thomson, J. Chem. Soc. p. 1118.
669 P. Trunel, Ann. chim. (11) 12, 93.
670 S. Umezawa, Bull. Chem. Soc. Japan 14, 363.
671 S. Uyeo and K. Higashi, J. Chem. Soc. Japan 60, 199.
672 "   "   "   " ibid. 60, 204.
673 L. Van Blaricom and E. C. Gilbert, J. Am. Chem. Soc. 61, 3238.
674 W. C. Vaughn, Phil. Mag. (7) 27, 669.
675 M. Yasumi, J. Chem. Soc. Japan 60, 1208.

1940

676 A. L. Bernoulli and H. Stauffer, Helv. Chim. Acta 23, 615.
677 C. S. Brooks and M. E. Hobbs, J. Am. Chem. Soc. 62, 2851.
678 E. Hertel and F. Lebok, Z. physik. Chem. B 47, 315.
679 M. E. Hobbs, J. W. Jacokes and P. M. Gross, Rev. Sci. Instr. 11, 126.

BIBLIOGRAPHY

<u>1940</u> (Cont'd.)

680   W. Hückel, J. Datow and E. Simmersbach, Z. physik. Chem. A <u>186</u>, 129.
681   G. Karagunis and T. Jannakopoulos, ibid. B <u>47</u>, 343.
682   J. A. A. Ketelaar, Rec. trav. chim. <u>59</u>, 757.
683   A. Kotera and Y. Go, J. Chem. Soc. Japan <u>61</u>, 455.
684   W. D. Kumler, J. Am. Chem. Soc. <u>62</u>, 3292.
685   S. Lee, J. Soc. Chem. Ind., Japan <u>43</u> Suppl., 190.
686   G. L. Lewis, P. F. Oesper and C. P. Smyth, J. Am. Chem. Soc. <u>62</u>, 3243.
687   " "   "   and C. P. Smyth, ibid. <u>62</u>, 1529.
688   R. Linke, Z. physik. Chem. B <u>46</u>, 261.
689   E. P. Linton, J. Am. Chem. Soc. <u>62</u>, 1945.
690   L. Malatesta, Gazz. chim. ital. <u>70</u>, 541.
691   "   "   ibid. <u>70</u>, 734.
692   A. A. Maryott, M. E. Hobbs and P. M. Gross, J. Am. Chem. Soc. <u>62</u>, 2320.
693   G. Mikhailov and D. Tischenko, Acta Physicochim. U.R.S.S. <u>12</u>, 129.
694   M. Milone and G. Tappi, Atti accad. sci. Torino <u>75</u>, I, 454.
695   H. Mohler and J. Sorge, Helv. Chim. Acta <u>23</u>, 1200.
696   Y. R. Naves and E. Perrottet, ibid. <u>23</u>, 912.
697   J. L. Oncley, J. Phys. Chem. <u>44</u>, 1103.
698   G. R. Paranjpe and P. Y. Deshpande, J. Univ. Bombay <u>9</u>, Pt. 3, 24.
699   H. A. Pohl, M. E. Hobbs and P. M. Gross, Ann. N. Y. Acad. Sci. <u>40</u>, 389.
700   I. A. Sheka, Zapiski Inst. Khim., Akad. Nauk U.R.S.R. <u>7</u>, No. 1, 57.
701   C. P. Smyth and G. L. Lewis, J. Am. Chem. Soc. <u>62</u>, 721.
702   " "   " , A. J. Grossman and S. R. Ginsburg, ibid. <u>62</u>, 192.
703   " "   ":, G. L. Lewis, A. J. Grossman and F. B. Jennings, ibid. <u>62</u>, 1219.
704   G. Tappi and C. Springer, Gazz. chim. ital. <u>70</u>, 190.
705   M. P. Volarovich and N. N. Stepanenko, Acta Physicochim. U.R.S.S. <u>13</u>, 647.
706   " "   "   " " "   "   J. Exptl. Theoret. Phys. (U.S.S.R.).
           <u>10</u>, 817.
707   Y. L. Wang, Z. physik. Chem. B <u>45</u>, 323.
708   A. Weizmann, Trans. Far. Soc. <u>36</u>, 329.
709   "   "   ibid. <u>36</u>, 978.
710   S. Winstein and R. E. Wood, J. Am. Chem. Soc. <u>62</u>, 548.

<u>1941</u>

711   A. Audsley and F. R. Goss, J. Chem. Soc. p. 864.
712   G. I. M. Bloom and L. E. Sutton, ibid. p. 727.
713   R. C. L. Bosworth, J. Proc. Roy. Soc. N. S. Wales <u>74</u>, 538.
714   L. G. S. Brooker, F. L. White, G. H. Keyes, C. P. Smyth and P. F. Oesper,
           J. Am. Chem. Soc. <u>63</u>, 3192.
715   B. C. Curran, ibid. <u>63</u>, 1470.
716   J. D. Ferry and J. L. Oncley, ibid. <u>63</u>, 272.
717   H. M. Foley and H. M. Randall, Phys. Rev. <u>59</u>, 171.
718   E. N. Gur'yanowa, Acta Physicochim. U.R.S.S. <u>14</u>, 154.
719   "   "   "   J. Phys. Chem. (U.S.S.R.) <u>15</u>, 142.
720   R. Heinze, M. Marder, K. H. Döring and K. Blechstein, Oel u. Kohle <u>37</u>, 8.
721   J. Henrion, Bull. soc. roy. sci. Liége <u>10</u>, 414.
722   E. Hertel and M. Schinzel, Z. physik. Chem. B <u>48</u>, 289.
723   K. Higashi, Bull. Inst. Phys.-Chem. Res. (Tokyo) <u>20</u>, 218.
724   "   "   Sci. Papers Inst. Phys.-Chem. Res. (Tokyo) <u>38</u>, 331.
725   "   "   and S. Uyeo, J. Chem. Soc. Japan <u>62</u>, 396.
726   "   "   " "   "   ibid. <u>62</u>, 400.
727   W. Hückel and W. Jahnentz, Ber. <u>74</u> B, 652.
728   K. A. Jensen and N. H. Bang, Ann. <u>548</u>, 95.
729   " "   "   " " "   "   ibid. <u>548</u>, 106.

1941 (Cont'd.)

730 K. A. Jensen and A. Berg, Ann. 548, 110.
731 Kojima, J. Chem. Soc. Japan 62, 903.
732 W. D. Kumler and I. F. Halverstadt, J. Am. Chem. Soc. 63, 2182.
733 E. Liese in E. Hertel, Z. Elektrochem. 47, 813.
734 A. A. Maryott, J. Am. Chem. Soc. 63, 3079.
735 " "   "   M. E. Hobbs and P. M. Gross, ibid. 63, 659.
736 Y. R. Naves and E. Perrottet, Helv. Chim. Acta 24, 3.
737 G. R. Paranjpe and D. J. Davar, Indian J. Physics 15, 173.
738 N. L. Phalnikar, B. V. Bhide and K. S. Nargund, J. Univ. Bombay 10, Pt. 3, 48.
739 C. P. Smyth, J. Am. Chem. Soc. 63, 57.
740 " "   "   J. Org. Chem. 6, 421.
741 G. Tappi, Gazz. chim. ital. 71, 111.
742 K. S. Topchiev, M. M. Yakshin and R. E. Shindel, Compt. rend. acad. sci. U.R.S.S. 30, 502.
743 Y. Tsudzuki and K. Higashi, Sci. Papers Inst. Phys.-Chem. Res. (Tokyo), 39, 185.
744 H. Ulich and G. Heyne, Z. physik. Chem. B 49, 284.
745 S. Uyeo, Bull. Chem. Soc. Japan 16, 177.
746 G. Venturello, Atti accad. sci. Torino, Classe sci. fis. math. nat. 77, I, 57.
747 W. Wassiliew and J. Syrkin, Acta Physicochim. U.R.S.S. 14, 414; J. Phys. Chem. U.S.S.R. 15, 254.
748 R. H. Wiswall, Jr., and C. P. Smyth, J. Chem. Physics 9, 356.

1942

749 A. E. van Arkel, P. Meerburg and C. R. v.d. Handel, Rec. trav. chim. 61, 767.
750 A. Audsley and F. R. Goss, J. Chem. Soc. p. 358.
751 "   "   " " "   " ibid. p. 497.
752 G. N. Bhattacharyya, Indian J. Physics 16, 369.
753 G. E. Coates and L. E. Sutton, J. Chem. Soc. p. 567.
754 B. C. Curran, J. Am. Chem. Soc. 64, 830.
755 N. R. Davidson and L. E. Sutton, J. Chem. Soc. p. 565.
756 I. F. Halverstadt and W. D. Kumler, J. Am. Chem. Soc. 64, 1982.
757 E. C. Hurdis and C. P. Smyth, ibid. 64, 2212.
758 " "   "   " " "   " ibid. 64, 2829.
759 K. A. Jensen and A. Friediger, Dansk. Tids. Farm. 16, 280.
760 A. Kotera, J. Chem. Soc. Japan 63, 364.
761 W. D. Kumler, J. Am. Chem. Soc. 64, 1948.
762 " "   "   ibid. 64, 2993.
763 " "   "   and G. M. Fohlen, ibid. 64, 1944.
764 " "   "   and I. F. Halverstadt, ibid. 64, 1941.
765 T. J. Lane, P. A. McCusker and B. C. Curran, ibid. 64, 2076.
766 L. Malatesta and R. Pizzotti, Gazz. chim. ital. 72, 491.
767 P. A. McCusker and B. C. Curran, J. Am. Chem. Soc. 64, 614.
768 P. F. Oesper, G. L. Lewis and C. P. Smyth, ibid. 64, 1130.
769 " "   "   and C. P. Smyth, ibid. 64, 173.
770 " "   "   " " "   " ibid. 64, 768.
771 " "   "   " " "   " and M. S. Kharasch, ibid. 64, 937.
772 W. T. Olson and F. M. Whitacre, Anesthesia and Analgesia 21, 106.
773 N. L. Phalnikar, J. Univ. Bombay 11, Pt. 3, 87.
774 I. A. Sheka, J. Phys. Chem. (U.S.S.R.) 16, 99.
775 J. Turkevich, P. F. Oesper and C. P. Smyth, J. Am. Chem. Soc. 64, 1179.
776 J. W. Zwartsenberg and J. A. A. Ketelaar, Rec. trav. chim. 61, 877.

1943

777   W. F. Anzilotti and B. C. Curran, J. Am. Chem. Soc. 65, 607.
778   H. J. Backer and W. G. Perdok, Rec. trav. chim. 62, 533.
779   R. Davis, H. S. Bridge and W. J. Svirbely, J. Am. Chem. Soc. 65, 857.
780   S. D. Gokhale, N. L. Phalnikar and S. D. Bhawe, J. Univ. Bombay 11 A,
        Pt. 5, 56.
781   N. B. Hannay and C. P. Smyth, J. Am. Chem. Soc. 65, 1931.
782   M. E. Hobbs and A. J. Weith, Jr., ibid. 65, 967.
783   E. C. Hurdis and C. P. Smyth, ibid. 65, 89.
784   K. A. Jensen, Z. anorg. allgem. Chemie 250, 245.
785   " "   "   ibid. 250, 257.
786   " "   "   ibid. 250, 268.
787   " "   "   and A. Friediger, Kgl. Danske Videnskab. Selskab, Math.-
        fys. Medd. 20, No. 20.
788   J. A. A. Ketelaar, Rec. trav. chim. 62, 289.
789   A. W. Laubengayer and G. R. Finlay, J. Am. Chem. Soc. 65, 884.
790   L. Malatesta and R. Pizzotti, Gazz. chim. ital. 73, 143.
791   "   "   " "   "   ibid. 73, 344.
792   "   "   " "   "   ibid. 73, 349.
793   J. L. Oncley, Chapter 22, Cohn and Edsall's Proteins and Amino Acids,
        Reinhold, New York, N. Y.
794   J. W. Zwartsenberg, Rec. trav. chim. 62, 148.

1944

795   W. Hückel and W. Wenzke, Z. physik. Chem. A. 193, 132.
796   T. L. Jacobs, J. D. Roberts and W. G. MacMillan, J. Am. Chem. Soc. 66,
        656.
797   E. F. J. Janetzky and M. C. Lebret, Rec. trav. chim. 63, 123.
798   A. Kotera, S. Nishimura and Y. Oto, J. Chem. Soc. Japan 65, 527.
799   V. Y. Krasil'nikov, J. Phys. Chem. (U.S.S.R.) 18, 174.
800   J. J. Lander and W. J. Svirbely, J. Am. Chem. Soc. 66, 235.
801   A. A. Maryott and S. F. Acree, J. Res. U. S. Nat'l. Bur. Stand. 33, 71.
802   A. Nikuradse and A. Berger, Physik. Z. 45, 71.
803   T. M. Shaw, E. F. Jansen and H. Lineweaver, J. Chem. Physics 12, 439.
804   N. Stepanenko, J. Exptl. Theoret. Phys. (U.S.S.R.) 14, 163.
805   "   "   and V. Agranat, ibid. 14, 226.
806   J. K. Syrkin and E. A. Shott-L'vova, Acta Physicochim. U.R.S.S. 19, 379.
807   G. Thomson, J. Chem. Soc. p. 404.
808   M. Volarovich and N. Stepanenko, J. Exptl. Theoret. Phys. (U.S.S.R.) 14,
        313.

1945

809   B. C. Curran, J. Am. Chem. Soc. 67, 1835.
810   F. Fairbrother, J. Chem. Soc. p. 503.
811   W. D. Kumler, J. Am. Chem. Soc. 67, 1901.
812   " "   "   and G. M. Fohlen, ibid. 67, 437.
813   J. J. Lander and W. J. Svirbely, ibid. 67, 322.
814   D. G. Leis and B. C. Curran, ibid. 67, 79.
815   S. R. Phadke, S. D. Gokhale, N. L. Phalnikar and B. V. Bhide, J. Indian
        Chem. Soc. 22, 235.
816   S. R. Phadke, N. L. Phalnikar and B. V. Bhide, ibid. 22, 239.
817   G. M. Phillips, J. S. Hunter and L. E. Sutton, J. Chem. Soc. p. 146,
818   W. C. Schneider, W. C. Carter, M. Magat and C. P. Smyth, J. Am. Chem.
        Soc. 67, 959.

<u>1945</u> (Cont'd.)

819   N. Stepanenko, B. Agranat and T. Novikova, Acta Physicochim. U.R.S.S. <u>20</u>, 923.

820   W. J. Svirbely and J. J. Lander, J. Am. Chem. Soc. <u>67</u>, 2189.

821   J. Syrkin and E. A. Shott-L'vova, Acta Physicochim. U.R.S.S. <u>20</u>, 397.

822   A. Weissberger, J. Am. Chem. Soc. <u>67</u>, 778.

<u>1946</u>

823   E. B. Baker, A. J. Barry and M. J. Hunter, Ind. Eng. Chem. <u>38</u>, 1117.

824   G. E. Coates, J. Chem. Soc. p. 838.

825   T. W. Dakin, W. E. Good and D. K. Coles, Phys. Rev. <u>70</u>, 560.

826   H. G. Emblem and C. A. McDowell, J. Chem. Soc., p. 641.

827   J. Granier, C. r. acad. sci. <u>223</u>, 893.

828   N. B. Hannay and C. P. Smyth, J. Am. Chem. Soc. <u>68</u>, 171.

829   "   "   "   "   "   "   "   ibid. <u>68</u>, 244.

830   "   "   "   "   "   "   "   ibid. <u>68</u>, 1005.

831   "   "   "   "   "   "   "   ibid. <u>68</u>, 1357.

832   O. Hassel and S. Ore, Tids. Kjemi Bergvesen Met. <u>6</u>, 72.

833a   K. Højendahl, Kgl. Danske Videnskab. Selskab, Mat.-fys. Medd. <u>24</u>, No. 2.

833b   "   "   and K. K. Møller, Dansk. Tids. Farm. <u>20</u>, 197.

834   N. C. C. Li and F-K. Chen, J. Chinese Chem. Soc. <u>13</u>, 1.

835   "   "   "   ", ibid. <u>13</u>, 8.

836   "   "   "   "   ibid. <u>13</u>, 11.

837   L. Malatesta, Gazz. chim. ital. <u>76</u>, 182.

838   B. Melander, Svensk Kem. Tid. <u>58</u>, 231.

839   J. G. Miller and H. S. Angel, J. Am. Chem. Soc. <u>68</u>, 2358.

840   G. R. Paranjpe and M. B. Vajifdar, Indian J. Phys. <u>20</u>, 197.

841   M. T. Rogers and J. D. Roberts, J. Am. Chem. Soc. <u>68</u>, 843.

842   "   "   "   "   A. Young, ibid. <u>68</u>, 2748.

843   R. O. Sauer and D. J. Mead, ibid. <u>68</u>, 1794.

844   B. I. Spinrad, ibid. <u>68</u>, 617.

845   O. J. Sweeting and J. R. Johnson, ibid. <u>68</u>, 1057.

846   H. Wild, Helv. chim. Acta <u>29</u>, 497.

<u>1947</u>

847   C. Curran and E. P. Chaput, J. Am. Chem. Soc. <u>69</u>, 1134.

848   E. N. Gur'yanova, J. Phys. Chem. U.S.S.R. <u>21</u>, 411.

849   "   "   "   ibid. <u>21</u>, 633.

850   V. Hardung, Helv. phys. Acta <u>20</u>, 459.

851   J. G. Jelatis, Tech. Rep. <u>7</u>, O.N.R. Contract N5 ori-78 T.O.1. Lab. Ins. Res., Mass. Inst. Tech.

852   R. J. W. Le Fèvre and P. Russell, Trans. Far. Soc. <u>43</u>, 374.

853   N. C. C. Li and T-L. Chu, J. Am. Chem. Soc. <u>69</u>, 558.

854   "   "   "   " , C. V. An and W. H. Wu, J. Am. Chem. Soc. <u>69</u>, 2558.

855   H. Lumbroso, C. r. acad. sci. <u>225</u>, 1003.

856   A. A. Maryott and S. F. Acree, J. Research Nat'l. Bur. Standards <u>38</u>, 505.

857   M. T. Rogers, J. Am. Chem. Soc. <u>69</u>, 457.

858   "   "   "   ibid. <u>69</u>, 1243.

859   "   "   "   ibid. <u>69</u>, 2544.

860   "   "   "   and T. W. Campbell, ibid. 69, 2039.

861   A. N. Shidlovskaya and Y. K. Syrkin, Compt. rend. acad. sci. U.R.S.S. 55, 231.

862   N. N. Stepanenko, B. A. Agranat and V. F. Yakovlev, J. Phys. Chem. U.S.S.R. <u>21</u>, 893.

BIBLIOGRAPHY

1948

863   T. W. Campbell and M. T. Rogers, J. Am. Chem. Soc. 70, 1029.
864   S. Golden, T. Wentink, Jr., R. Hillger and M. W. P. Strandberg, Phys. Rev.
      73, 92.
865   N. C. C. Li and T. D. Terry, J. Am. Chem. Soc. 70, 344.
866   G. Potapenko and D. Wheeler, Jr., Rev. Mod. Phys. 20, 143.
867   W. C. Schneider, J. Am. Chem. Soc. 70, 627.
868   W. F. Seyer and G. M. Barrow, ibid. 70, 802.
869   M. W. P. Strandberg, T. Wentink, Jr. and R. L. Kyhl, Tech. Rep. 59, Res.
      Lab. Electronics, Mass. Inst. Tech.
870   A. J. Weith, M. E. Hobbs and P. M. Gross, ibid. 70, 805.
871   T. Wentink, Jr., M. W. P. Strandberg and R. Hillger, Bull. Am. Phys. Soc.
      23, No. 2, p. 18.